D0855219

ECO-SOCIALISM

Capitalism continues to degrade ecosystems and create social injustice. The 1992 Earth Summit demonstrated that the powerful vested interests behind Western capitalism have no intention of radically changing their goals and methods to help create an environmentally sound or socially just global society. In order to confront this, the green movement must now develop a coherent eco-socialist politics. People must control their own lives and their relationship with their environment.

Drawing on Marx, Morris, Kropotkin and anarcho-syndicalism, David Pepper presents a provocatively anthropocentric analysis of the way forward for green politics and environmental movements. Establishing the elements of a radical eco-socialism, the book rejects biocentrism, simplistic limits to growth and overpopulation theses, whilst exposing the deficiencies and contradictions in green approaches to postmodern politics and deep ecology.

Eco-Socialism will provide students of ecology, politics and the environment with a thorough introduction to the ideologies of Marxism, anarchism and deep ecology, and how these can be synthesised into a radical green politics.

David Pepper is principal lecturer in Geography at Oxford Brookes University.

ECO-SOCIALISM

From deep ecology
to social justice

David Pepper

London and New York

First published 1993
by Routledge
11 New Fetter Lane, London EC4P 4EE

Simultaneously published in the USA and Canada
by Routledge
29 West 35th Street, New York, NY 10001

© David Pepper

Typeset in Garamond by Witwell Ltd, Southport
Printed and bound in Great Britain by
Mackays of Chatham PLC, Chatham, Kent

All rights reserved. No part of this book may be reprinted or
reproduced or utilized in any form or by any electronic,
mechanical, or other means, now known or hereafter invented,
including photocopying and recording, or in any information
storage or retrieval system, without permission in writing from
the publishers.

British Library Cataloguing in Publication Data
A catalogue reference for this book is available from the British Library

Library of Congress Cataloging-in-Publication Data
Pepper, David.
Eco-socialism: from deep ecology to social justice/David
Pepper
p. cm.
Includes bibliographical references and index.
ISBN 0-415-09718-5 (hb.) — ISBN 0-415-09719-3 (pbk.)
1. Human ecology—Political aspects. 2. Social ecology. 3. Deep
ecology. 4. Green movement. I. Title.
GF21.P4 1993
304.2′8—dc20 93-16565
CIP

To Nickie

CONTENTS

TABLES AND FIGURE

FOREWORD

I finished the penultimate draft of this book just as the 1992 Earth Summit in Rio de Janeiro closed. Jonathon Porritt, that most public of British green spokespeople, declared that he had gone to the summit with low expectations and had them all met! This book sets out, among other things, some reasons why Porritt was indeed wise to have low expectations.

Many other greens, however, declared their disappointment at the Summit's meagre outcomes. This must mean that they somehow expected the world's richest nations to sacrifice a substantial part of their riches and, more significantly, the means of obtaining them, to help the poorest nations to protect the environments which they now have to destroy in order to survive and develop in the world economic system. We should all, however, appreciate that being capitalist nations, the USA, the EC, Japan and the like *cannot* do this in any serious and permanent way without ceasing to be what they are. Marxist analysis reveals why this is so, and it also suggests how best to think about change towards radically alternative economic and social arrangements, of the kind which the concept of a truly commun(al)ist 'sustainable development' demands.

There are many other things about Marxism which greens may find useful and interesting, and I attempt to outline them here. I also describe the influence which anarchism has had on present green political philosophy, and I suggest what elements of this influence should be retained and what should be discouraged. The aim is to outline an *eco-socialist* analysis that offers a radical, socially just, environmentally benign – but fundamentally anthropocentric – perspective on green issues.

For I think that this is what the green movement now needs, rather than its current 'biocentric' and politically diffuse approach, in order to appeal to the concerns of the many who are still alienated by or indifferent to it. Furthermore, and pragmatism aside, I think it important not to allow our concern for non-human nature to become a substitute for, or a priority over, concern about people. Some greens believe that we should protect and respect nature for its 'intrinsic worth', whatever that is, rather than its worth for (all) people. I am not comfortable about this. Social justice, I think, or the increasingly global lack of it, is the most pressing of all environmental problems. And the Summit showed clearly that attaining more social justice is the *prerequisite* for combating ozone depletion, global warming and the rest.

All this is the political message of the book. However preaching is not its main purpose. That is to explain as lucidly as possible what Marxism and anarchism are about, and what is their relevance to some of the most pressing political issues which the green movement raises.

It is mainly intended for students in various disciplines, and for all interested but not particularly academic people in and around the green movement. It aims to synthesise and represent clearly the views of Marxists, anarchists and others who may not have written primarily for such an audience. It arose out of my attempts to prepare a substantially revised new edition of *Roots of Modern Environmentalism*. Having surveyed my profuse notes, collected since the first edition was published, I surmised that I would probably need about a third of a million words to say all that I needed to, and that (quite unreasonably) the publisher would not let me have them! This, then, constitutes my further reflections on just the sixth and seventh chapters of that book.

It does not set out to achieve the same breadth or scope as *Roots*. For one thing, it assumes that readers already know something of the concerns and approaches of ecocentrism (as set out briefly in *Roots* or very fully in Andrew Dobson's excellent book on *Green Political Thought*). For another, it does not intend to chart comprehensively all the possible roots of political ecology and ecological politics (other recent books have done this), but to concentrate on Marxism and anarchism. And within these boundaries there are further limitations. For instance the Marxist economic theory is sketched out in its basics only, although I concede that a major and urgent task of eco-socialism is to grasp the nettle that the green movement has often avoided and get to grips with the *detail*s of a green socialist political economy. And, in discussing agents and actors in radical eco-socialist change, I assert the continuing importance of a (world) proletariat, but do not get round to the also-important issue of where the self-employed or the managerial classes might fit in any collective radical movement. Furthermore the book is overwhelmingly about theory, whereas a much slimmer volume with feasible suggestions about what to *do* might contribute so much more. And where, I hear readers ask, is the discussion of 'a feminist perspective' which is becoming almost mandatory for this sort of book?

I confess that I would like the book not to have these and other shortcomings, which are certainly not the fault of those whose help I acknowledge below. Out of the range of possible excuses, I am not sure which to select. Insufficient space? I think I prefer the one about merely intending to suggest *some possible* items on an *agenda* for future discussion. However, the truth is that I have still got a lot to learn and think out. But then so has everyone else in and around the green movement.

I learned something through writing *Roots*, which was better received and more widely used than I had expected. But some green critics hated its Marxist leanings. I hope that I can win them over by this fuller and, I hope, more satisfying account, and that it will help us all to sharpen up our thinking – and our act – in the face of the continuing stubborn refusal of the green millennium to appear over the horizon (the 'new world order' having shown itself to be merely a new order of exploitation of people and nature).

I thank the following for their help, either in supplying material, or in criticising part of the text or simply in encouraging me to think critically about particular ideas which have subsequently featured in the book: Adam Buick, Dave Elliot, Nickie Hallam, Jim O'Connor, Phil O'Keefe, Chris Park, Richard Peet, Graham Purchase, Biff Shore and Frank Webster. And I am glad to have been able to listen to many recordings of the discussion meetings of the Socialist Party of Great Britain. I have found them informative and challenging; the people who make them available do a considerable service, and I recommend them to readers who want to find out about socialism from socialists rather than just from more detached and less exciting academic textbooks.

The latter part of Chapter 4.4 originally appeared as an article in *The Raven*, 1(4), March 1988, and Chapter 2.3 is taken from an article forthcoming in the journal *Capitalism, Nature, Socialism*.

David Pepper, Oxford 1992

1

RED AND GREEN: OLD OR NEW POLITICS?

1.1 THE RED-GREEN DEBATE

Ten years ago, a friend asked me to address a local Friends of the Earth meeting which he was organising. My interest was in the historical roots of green philosophy, so I regaled my audience with accounts of William Morris, Peter Kropotkin and the like. Naively, I mentally prepared myself to receive accolades in the ensuing discussion for drawing such historical links. Instead I sensed antipathy mingled with hostility from some. They were disappointed. Had I not realised that what the greens were saying had never been said before? Did I not appreciate its distinctiveness from conventional politics?

I had offended a fundamental aspect of green psyche, which holds that ecologism really is about a new world order, and a new 'politics of life' (to use the Green Party slogan). I compounded this crime by suggesting, in *Roots of Modern Environmentalism*, that greens needed to assimilate Marxist perspectives into their analysis. This was a red rag in the face of a green bull, being dismissed as 'just so much angry spluttering from worn-out ideologues who have lost touch with the real world' (Porritt and Winner 1988, 256).

Notwithstanding this familiar criticism, I, a clapped-out ideologue and aspiring member of that 'malign force', the Marxist left (Porritt and Winner, p. 220), intend to splutter on unabashed. I will try, in the following pages, to extend and deepen the recent debate between the red and green positions on our 'ecological crisis'. This is because I do not accept Adrian Atkinson's dismissal of this debate as a mere 'argument' between two views that, in practice, display no fundamental contradiction.

True, there are many conjunctions between red-greens and green-greens (these terms both describe *radical* ecologists, or 'ecocentrics' as opposed to 'light' greens or 'environmentalists', i.e. technocentrics, who are not the subject of this book – see Chapter 2). If red-greens make much use of Marxism, however, green-greens are more indebted to anarchism. And although the two often conflate in the anarcho-communism of the likes of Kropotkin, elements of which form a template for modern ecotopias, and for the social ecology of

1

Murray Bookchin, there are also significant – potentially irreconcilable – differences between them.

This is important, in these days of tentative radical alliances and red–green networks, for reasons which Tony Benn gave (cited in Porritt and Winner, p. 69):

> Until the basic principles of socialism are re-established (equity, democracy, accountability, internationalism and morality), one cannot build non-opportunistic, genuine relations with movements which are themselves divided over the primacy of these principles.

I think it is time we had the whole thing out, and this book intends to contribute to that process – a process which is of more than just academic importance. For Western capitalism is yet again in crisis, and more than ever before the effects of the crisis extend across the world. At the same time that recession and retrenchment have decimated manufacturing industry in the old heartlands, and people stubbornly refuse to consume their way out of slump, capitalism's response has been to reach ever deeper into second and third worlds for markets and sources of cheap labour and materials.

The current search for a new, more 'liberal' General Agreement on Tariffs and Trade (GATT) signifies an attempt to bring everyone unambiguously into the global capitalist economy. This threatens a further mushrooming of what neither socialists nor greens want – the hedonistic consumer society with a high throughput of goods but a low output of human fulfilment. In it, disenfranchised and underprivileged groups are increasingly economically marginalised and the environmental costs of the search for profits mount. But these twin evils of social injustice and environmental degradation will continue to grow, even though most people recognise them as evils, for there is no prospect that their present root causes in the economics and politics of capitalism will be radically examined and tackled. The 1992 Earth Summit in Rio de Janeiro made this plain. For while some third world leaders and other eminent public figures correctly identified the problems and their causes, Western leaders staunchly defended the 'right' of multinational capital to continue operating in the same old way and resurrected old Malthusian (third world) 'overpopulation' canards for their explanations of causes. Faced with draft global accords, conventions and other agreements to take fundamental action on social and environmental problems they watered them down, prevaricated and even refused point blank to sign them. Or, more dishonestly, they did so and then went home and carried on with the same old policies.

It is at times like these that the left and the greens anticipate that they will make their mark most effectively. Yet this has not happened. The almost world-wide disarray of the left in the 1980s is well documented. But the greens, who promised us a 'new politics' to replace both socialism and capitalism, have also been on the retreat. Electoral gains made in Europe in the early 1980s have been substantially relinquished – indeed the British Green Party faces crisis at the very time of writing, with the resignation of half its executive council and a fall in membership from nearly 20,000 in 1990 to 10,000 in 1992 (*The Times*, 28 August 1992).

Since, then, the pragmatic, 'realist', but anodyne politics of social democracy, democratic socialism and green reformism have failed to mount a serious threat to the status quo; more radical socialists and greens argue afresh that what is needed is the much more fundamental politics of *eco-socialism*. And yet, for all the exploratory red–green 'networking' that goes on, no very potent, effective and coherent eco-socialism has emerged. I think that this is because of the fact that to bring together red and green you have, effectively, to unite socialism with anarchism – the traditional political philosophy which more than any other informs the green movement. This is not as easy as it sounds, because, contrary to popular misconception, it is not always possible to regard anarchism as just another form of socialism. This book tries to help the cause of eco-socialist politics by describing and explaining the forms of socialism – particularly Marxist socialism – and anarchism on which they must be based. It highlights and clarifies many of the differences between socialism and anarchism in order to suggest the agenda for any future political discourse which wants to close the gap and create eco-socialism as a more vital force. It does this by suggesting that greens should make more of an accommodation with reds by dropping those aspects of their anarchism that are more akin to liberal and postmodern politics. At the same time reds should accommodate with greens by reviving those traditions in socialism which I describe and review here – including traditions of decentralism and of the society–nature dialectic, along with some resuscitation of orthodox Marxism's materialism and emphasis on rediscovering our power as producers.

The main part of the book maintains, in Chapter 3, that Marxist perspectives have more to offer greens than just an incisive analysis of capitalism, important as this is. Marxism suggests a dialectical view of the society–nature relationship, which is not like that of ecocentrics or technocentrics, and challenges both of them. It has a historical materialist approach to social change, which ought to inform green strategy. And it is committed to socialism, as Benn defines it above. And, yes, it is, and I am, anthropocentric enough to insist that nature's rights (biological egalitarianism) are meaningless without human rights (socialism). Eco-socialism says that we should proceed to ecology from social justice and not the other way around.

Many greens (e.g. Schumacher 1973) have said that Marxism is rigid, inflexible, deterministic, mechanistic (rather than organic), overly 'scientific' (in the positivist sense) about history, lacking humanism and a spiritual dimension, a 'bible' consisting of a set of prophecies which are mostly wrong, and totalitarian in outlook and implications.

What I have read about Marxism suggests that these criticisms are often partly or wholly inaccurate. What follows may illustrate this, although it is not intended as an apologia for Marxism's shortcomings. As Sarkar (1983, 164)

3

says: 'The point is not to find out the authentic Marx . . . the purpose is not to save Marxism, but to find out the truth . . .', and, citing Ullrich (1979, 95):

> it is now time that the senseless game of substituting endless quotations from the 'holy scriptures' for the analysis of new phenomena and one's own thinking is finally given up. . . . It is, moreover, unmarxist. Marx himself did not like to be called a Marxist. Today he would certainly not be a Marxist in the sense of uncritical adherence to the contents of his over one hundred years old writings.

My second contribution, in Chapter 4, is to outline the tenets of anarchism and how much they at present inform the position of what I call 'mainstream' greens (ecocentrics) as well as those who openly call themselves 'green anarchists'. While I will not argue in the conclusion (Chapter 5) that anarchism must be abandoned, I will highlight the distinctiveness of socialism and its debt to Marxism and suggest a shift in emphasis for ecocentrics towards this latter. Some greens may say that this shift is already occurring, but I wonder if they realise its full implications; such as possibly abandoning the idea of a money-driven economy, or that of biocentrism?

Before all this, I want to set the context of the debate, in Chapters 1 and 2. Some academics, like Atkinson (1991), Bramwell (1989) or Dobson (1990, 205–6) maintain that ecologism is 'a political ideology in its own right' because 'the descriptive and prescriptive elements in the political ecology programme cannot be accommodated within other political ideologies (such as socialism) without substantially changing them . . .'. For Dobson, this distinctiveness hinges particularly upon ecologism's acceptance of limits to growth and on the bioethic (advocating respect and reverence for the intrinsic value of 'non-human' nature – in its own right and regardless of its usefulness to humans). For Atkinson (p. 19) it is ecologism's utopianism (after, particularly, utopian socialists) which makes it

> a coherent political paradigm quite distinct from the conservatism, liberalism and socialism which today are commonly seen as defining the limits of the political spectrum.

All of this is arguable. For one, few greens nowadays propose *no* forms of economic growth for the future, while the argument itself that 'resources' are finite is intellectually problematic (see Chapter 3.5). Secondly, there are all sorts of objections to intrinsic value theory for nature – its theoretical and practical implications, its indebtedness to intuition rather than rational argument, its *impossibility* (we cannot know if nature values *itself*: we, as humans, can only approach nature from an anthropocentric standpoint) (Fox 1990, 184–96) and its tendency to set up a society–nature dualism (see Chapter 3.6). Thirdly, to suggest that modern politics have no utopianism may be true in the narrow sense; but their roots do. Marxism and anarchism themselves are utopian in the sense of having a vision of at least the principles of an ideal

4

(anarcho-socialist) society. But the former is not utopian in terms of how we go about *changing* society, and it justly criticises anarchists, utopian socialists and greens for being so (Chapter 3.9).

However, I do not want to pursue these objections here, and I do want to concede that the green political claim to distinctiveness, even newness, in its descriptive elements, may be accurate. Nonetheless, I consider that in their *prescriptive* elements: in how they propose to change and organise society, then they are often rehashing some old solutions to some very old and basic political questions. There is nothing wrong in this, but the rehash does need to be coherent, and greens widely recognise that such coherence is presently lacking. I propose that some attention to the perspectives of Marxism could lend ecologism a coherence that is appropriate for a forward, not a backward looking politics. This, together with the progressive elements of anarchism, might present green socialism as a form of socialism which is less prone to totalitarianism than some previous 'socialisms', though it will still entail sacrifice of some extant liberal 'freedoms', as is recognised in the conclusion; but this may be no bad thing.

To illustrate and emphasise that

> The political meanings attributed to 'social ecology' or 'the ecological paradigm' really derive from, and can only be discussed in terms of, traditions and debates (individualism versus collectivism, competition versus mutuality, authority and hierarchy versus liberty and equality) which long predate the emergence of ecology as a scientific discipline.
>
> (Ryle 1988, 12).

I shall begin by outlining briefly what some of these debates are about (see Table 1.1). They still largely set the fundamental political agenda for the twenty-first century, and the arrival of a green consciousness does not alter this: they form the context in which green politics are inescapably set. Marxism and anarchism have much to say about these debates.

It should be understood that the discussion in the following section, 1.2, is illustrative only. It does not purport to be an exhaustive list of all of the most important questions underlying the 'old' politics. Thus I do not debate in the abstract Ryle's authority/hierarchy versus liberty/equality dualism, or issues to do with technology (should it be 'hard' or 'soft', 'high' or 'appropriate', and does it determine social development or vice versa?) or scale (economies of scale versus small-is-beautiful) or whether the approach to politics should be reformist or radical. It may be argued that I should have done, for these questions figure centrally in modern discussions about ecology and so they inevitably figure in the anarchist as well as the Marxist discourses of Chapters 4 and 3. However, to limit the size of this chapter I have chosen some issues that do not so openly appear in green debates as such, but which, I think, ought to. I should also qualify the discussion by acknowledging that although much of it is presented in terms of conflicting dualisms, the issues are usually more

5

Table 1.1 Some fundamental social questions that underlie traditional and green political debates

Questions dealt with in Chapter 1:

HUMAN NATURE:	Is there such a thing? Is it shaped by environment or genetic inheritance? Is it greedy, aggressive and competitive or the reverse?
DETERMINISM or FREE WILL:	Are individuals and society the product of external forces – God, environment, economics – or do they have freedom to shape the world how they want it to be?
IDEALISM or MATERIALISM:	Are societies and economics shaped and changed as a result of new ideas and arguments which persuade people to act differently? Or are material, especially economic, structures and events, the main influences on behaviour and ideas?
INDIVIDUALISM or COLLECTIVISM:	Will social change be triggered by the actions of individuals changing their lifestyles and thoughts (often as consumers) or by groups taking collective action for political effect (often as producers)?
GEMEINSCHAFT or *GESELLSCHAFT*:	Is society merely a collection of individuals supporting each other for mutual gain? Or is there more to society than the sum of the parts – is it an entity to which individual interests are largely subordinate?
CONSENSUS or CONFLICT:	Which of these constitute the main motor of social change? Is society a genuine democracy whose state represents an equilibrium between the interests of all groups. Or is it dominated by elites (economic or otherwise) whose interests conflict with the majority?
STRUCTURALISM:	Are social events and individual or group behaviour (surface structure) a product of deep subconscious or hidden underlying structures in the human mind or in cultural or economic organisation? Or is what we easily recognise around us the full extent of social reality?
DEVELOPMENT:	Is the social and economic development of regions and nations best described by models of environmental determinism, or structural functionalism, or structural Marxism, or modernisation or dependent development or a mix of several of those? Which development model is most ecologically preferable – independent development (bioregionalism), or socialism?
EGALITARIANISM:	Should we support economic development that produces gross maldistribution of wealth and then put up with or mitigate these effects? Or should we promote a model which does not allow inequalities to develop in the first place? Should all living species be equally respected and treated?
'FREE MARKET' or INTERVENTION:	Which produces most benefit for most people? Can social and environmental need be met without planning and intervention in the free market? Do the latter stifle innovation and produce inefficiencies?

Some other questions, that arise in Chapters 3 and 4:

AUTHORITY or LIBERTARIANISM:	Is a peaceful, just, fulfilling and pleasant society more likely to result from one that is highly ordered and controlled through hierarchies dominated by the state or by elite groups? Or are the lack of hierarchies and a state, and the promotion of democratic self-organisation the keys to such a society?
SMALL or LARGE SCALE:	Is large-scale urban, industrial and political–economic organisation the most efficient way to achieve the desired society, or is small beautiful?
TECHNOLOGY – DETERMINED BY or DETERMINING:	Does society, or specific elements in it, control and determine technological developement? Or does the latter have a life of its own, which substantially influences the shape of society?
TECHNOLOGY – HIGH or INTERMEDIATE:	Which serves the interests of a socially just and ecologically sound society? Can the former express and be part of democratic social relationships? Can the latter provide large populations with basic needs?
MODERNISM – POSTMODERNISM:	Is the Enlightenment project of seeking universal good through understanding and establishing general rational principles (including an absolute morality) still feasible? Or should life be lived according to hedonistic principles celebrating the here and now, images rather than reality and the equal validity of all views and perspectives?

complicated than that. Some greens may object that the very process of polarising issues in this way is part of the problem rather than the cure. Dualistic thinking, they say, underlies the 'Enlightenment Project' (i.e. all those social and political ideals and goals which evolved in the period of capitalist development, scientific discovery and philosophical advance that occurred from the seventeenth century onwards). And it is this 'project' and that dualistic thinking which has ruined us – particularly the tendency to dualise society and nature, i.e. to see them as separate and opposite. There is much in such arguments, though they are not totally convincing. Nonetheless I have found dualistic thinking a very useful pedagogic device: we can often grow towards appreciating complex and multifaceted issues by first conceiving of them in simple – even simplistic – dualisms. They give us a toehold by which we can elevate ourselves eventually to a higher understanding of complex reality. Since the prime purpose of this book is to allow students and other interested people to become familiar with the debate, then the more I can assist their learning the better.

Having established that conventional political questions are not irrelevant in green concerns, I will go on to point out, in Chapter 2, that ecologism takes positions which draw on some traditional theories concerning political economy. I will also map out how ecologism might be seen in relation to other political ideologies, including Marxism.

7

All this means that I must reject Atkinson's startling contention (p. 43) that 'A consistent political ecology is not the negation of any particular European intellectual tradition but of the tradition as a whole' (despite his affirmation (p. 177) that the wholesale rejection of Marxism would be a disaster). Nor can I support the postmodernist rejection of the goals of the Enlightenment Project which he seems to flirt with, alongside so many greens. Agreed, those goals, of general human progress through rationality, science, industry and social justice, must have an ecological sanity infused into them. This cannot happen under capitalism, but I doubt also that it could occur within an autarkic (i.e. decentralised), bioregional development model where all ethics and economics, apart from those towards nature, were treated as totally relative and equally valid. However, a form of Marxist socialism which, it must be conceded, has constituted a minority tradition alongside 'actually-existing' socialisms, could provide many answers in the attempt to resolve an 'ecological crisis'. It could be the key to reshaping society radically while avoiding the loss for everyone of the many benefits that have been reaped for some during the capitalist phase of the Enlightenment Project.

1.2 SOME OLD POLITICAL QUESTIONS

Human nature

Is human nature aggressive or gentle, competitive or cooperative, selfish or giving? Any answer you get is almost certain not to be scientifically valid, being ostensibly a judgement about what most humans in time and space were and are fundamentally like, yet really based on observations drawn from a pitifully small sample of people. We can never properly substantiate a view of 'universal' human nature for this reason, and also because it seems impossible to separate innate characteristics ('nature') from those acquired from the environment ('nurture').

Hence the really important question is why so many people think that answers *can* be found and are significant. Arch English conservative Peregrine Worsthorne (1984) provides a clue in his defence of social hierarchies

> which developed in England over the centuries . . . [and] gave much quiet satisfaction from top to pretty well near the bottom, since a society where everybody knows their place is much more comfortable for all concerned.

'Hierarchy', he says, 'is not unpopular in itself since it is felt to be natural, which is to say *inevitable*' (emphasis added). Here, he uses the huge power behind the idea of nature and 'naturalness' as a *legitimator*. If what *I* do and like is natural, it is just, or must be accepted even if it is not liked. Conversely, if I do not like things *you* like – such as homosexuality or egalitarianism – I can dismiss their worth by branding them as 'unnatural'.

It works both ways. Differences between our political ideologies – what we believe and why – may well rest on our different feelings at heart about the nature of human nature (Goodwin 1982). Conversely, if I want to affirm my ideology over others I will try to show that it accords with 'human nature'.

Conservatism, particularly, is legitimated by the idea of nature and the natural order. Thomas Hobbes said: Men from their very birth, and naturally, scramble for everything they covet, and would have all the world, if they could, to fear and obey them. This justified Edmund Burke in advocating social control: 'the passions of individuals should be subjected and the inclinations of men should frequently be thwarted . . . by a power out of themselves'.

And 'naturalness' also justifies the irrational belief (of liberals as well as conservatives) that land should be owned as private property rather than being held in common: 'an absolute and irreducible need' instinctively 'rooted in nature' (Scruton 1980, 99 – all quotations in Coleman 1990, 8). Conservatism goes on to argue something that many greens hold dear: that nature is, or ought to be, a model for human society. Social Darwinism holds that Darwin's motor of evolution for animals and plants – competition and struggle for scarce resources leading to survival of the fittest thus enriching the whole species – can also propel human societies towards perfection. Hence the need to conserve, uninterfered with, the competitiveness and struggle of 'free market' liberal capitalism. Social Darwinists are generally oblivious to the circularity of their argument; for Darwin's evolutionary 'laws' were, self-confessedly, drawn in the first place from Thomas Malthus's and Herbert Spencer's observations of *human* society. Hence social Darwinism is really social Spencerism (Oldroyd 1980).

Modern sociobiology tends to argue this way too. Sociobiologists like Konrad Lorenz and Desmond Morris emphasise the innateness of aggression and competition, and suggest that behind the veneer of civilisation we are all selfish 'primitives'. Almost perversely, Richard Dawkins (1976) insists that even apparently altruistic acts really stem from *self*-interest, and then he is distressed when right-wing ideologues latch on to his theories.

Left-wing ideologues propose various positions. Among them is the view that nature *is*, indeed, a model for human society, but that nature is inherently *cooperative*. This was Kropotkin's celebrated theory of mutual aid (Chapter 4.3). Greens, like Capra, frequently espouse it too. Others, like Rose, Kamin and Lewontin (1984), rebut the scientific respectability of evidence for characteristics like intelligence being inherited, and argue for a Marxian dialectic between nature and nurture, in which each shapes the other.

This develops into an argument for the essential *socialness* of human nature. Human nature may therefore be moulded by moulding the social environment which produces it:

Any general character, from the best to the worst, from the most

9

ignorant to the most enlightened, may be given to any community and even to the world at large.

(from Robert Owen's *A New View of Society*, cited in Coleman 1990)

William Morris, while rejecting Owen's approach of setting up 'ideal communities', nonetheless adopted a Marxian perspective on human nature – as plastic, not finalised, and therefore a product of human history (Chapter 4.3).

Atkinson (1991, 69) encapsulates the real nature of the debate about human nature:

> The English common sense assumption concerning human nature and the organisation of society necessarily embodying hierarchical relations, that emerges constantly as the essential 'discovery' of British social theory [Hobbes, Hume and the economic theory of Adam Smith] is no more than cultural prejudice reinforced by . . . cultural prejudice.

Atkinson goes on to argue that human nature is *not* a barrier to social improvement – a traditional socialist view and also a crucial one to greens, who do want radically to improve society:

> Other societies are organised around different cultural assumptions and history demonstrates regularly that change in assumptions and organisational arrangements does occur and is possible.

The question about what human nature is 'really' like, then, is not the crucial one, compared to that which asks if it can feasibly be *changed*. For greens to spend much time on the former is at best a waste of time. For instance, whether we are basically cooperative or competitive is in a way a red herring. The apparent cut-throat competition of capitalist economics is really a highly *cooperative* affair. Exploiters and exploited have to agree to occupy their roles and to accept the goals of capitalism as economic and cultural norms: as witness how 'deeply cooperative [with the bosses] sentiments ran within the postwar [US] workforce' (Harvey 1990, 133). The important questions are about the purposes to which we devote our cooperative abilities.

And there are other blind alleys in this quest for human nature. For instance, greens persist in holding up aboriginal peoples as ecologically sound 'natural societies' (e.g. the American Indian). Yet this concept of the 'noble savage' is as ideological and subject to historical fashion as that of human nature itself. People tend to find what they want to find in such 'traditional societies' (BBC 1992).

Determinism and free will

Just how free are humans to control, collectively or individually, their lives, their social and economic arrangements and their relationship with nature? This is a crucial political as well as philosophical question. As with the idea of

10

human nature, that of limits on human action set by supposedly external forces – e.g 'laws' of economics or history, God's design, technological progress or the physical environment – can powerfully legitimate the status quo. To say that we are *determined* by outside forces is potentially to argue that change which is out of sympathy with such forces is ill-advised if not impossible. And it can also suggest that features of society which we do not like (unemployment for instance) must be suffered because they result from forces (economic laws, world recession) beyond the control of government.

It would thus seem to be against the interests of any group wanting radical social change to support deterministic (therefore perhaps fatalistic) arguments, rather than the idea that humans can freely shape their own society – 'make their own history' in Marx's phrase. And socialists do generally shun such arguments. Greens, however, have a tradition (Goldsmith *et al.* 1972, Ekins 1986, Ehrlich and Ehrlich 1990) of accepting environmental (resource) limits as immediately circumscribing and determining human activity: hence their strictures against economic and population growth.

This is a form of environmental determinism (as opposed to the biological determinism of genetic inheritance, i.e. 'human nature'). Environmental determinism has appeared in many guises, from the Malthusian limits to growth thesis, to that of early geographers (Peet 1985) that human nature, physiognomy and national and social characteristics are more or less determined by climate, soil, relief and geographical position (still a popular notion with many people). And the view that the *built* environment controls human character and nature has strongly featured in all attempts at social engineering, from utopian socialist communities to twentieth-century urban planning and architecture.

Cornucopian technocentrics and free-market advocates often reject the limits to growth thesis (Simon and Kahn 1984), emphasising the Baconian creed that scientific knowledge equals power over nature: a power which should be used to improve humankind's lot by extending the boundaries of nature's 'limits'. In a way their arguments are equally deterministic, suggesting that humans can determine nature's form and behaviour through adequate knowledge of cause-and-effect laws governing its various components and their relationships. But in another way they can be seen as supporting the idea of freedom of human will – freedom to control an external environment. Indeed, essential corollaries of such views are that nature *is* external to, or separate from, us, and that nature is like a machine (Pepper 1984, 46–54, 117–18). Both these ideas are apparent anathema to deep ecologists.

Less materialistic Western philosophies which also emphasise human freedom of will in relation to society and nature have been developed in the last hundred or so years as phenomenology and existentialism. The science of phenomenology assumes that we are not separate from the rest of the world and are not predetermined by 'external' forces. Indeed it emphasises the way *we* shape the world: imposing structure, meaning and value onto it via our consciousness. This is not to deny the existence of an 'objective' nature 'out there' (though more extreme idealist philosophies like that of George Berkeley and those of New Ageism did and do see matter purely as a manifestation of mental activity (Lacey 1986, 97)). But it is to suggest that this does not really matter (Warnock 1970, 26–8). *Important* knowledge of the world is knowledge of how the consciousness and intentions of individuals and groups interpret, mediate and indeed structure it. Since consciousness and perception vary between individuals and groups, this science therefore emphasises *subjective* ways of knowing the world, through intuitive understanding. Thus, how different people and cultural groups know and understand their own world of immediate experiences – their 'lifeworld' – is vital. This suggests a relativist view of knowledge, understanding and, indeed, ethics concerning how the world *should* be. It implies that the knowledges of different individuals and groups can be regarded as equally important and valid.

By extension, the individualist philosophy of existentialism says that there are no objective, external facts or laws governing our social existence, save that we are born and one day will die. We are not helpless playthings of historical forces, or social laws and codes of conduct. *We* have control and choice over most facets of our existence; not being bound by economic or social conventions. This is not to deny totally that our environment, including culture, society and economics, *conditions* our situation. But 'condition' does not mean 'determine'. So we must accept that while on one hand we have been thrown into a world which is not of our making, on the other we are free to decipher the meaning of that world for ourselves, not as interpreted by others or supposedly external factors beyond our control.

Not to recognise this is to lead an alienated and 'inauthentic' existence. But if we do recognise it we open up a horizon of possibilities, including people being made according to how they *desire* to be. This carries all sorts of implications for our relations with other people and nature. While it could be interpreted as a doctrine of selfish individualism, it does, however, argue that since we have been free to make our world, the world we experience – polluted, socially unjust – is our creation, for which we ultimately are therefore responsible.

Free-will philosophies have much political affinity with anarchism (Chapter 4.3), and they strike some chords with Marxist liberation theory (Chapter 3.7). But whereas they address the issue by exploring it in the realm of consciousness and ideas, Marxism is particularly concerned with how this realm relates to the material sphere, particularly that of production and economics. And while the focus of existentialism is the individual, Marxism is keen to emphasise how individuals are socially conditioned and materially bound (especially if they belong to the underclasses in society) and hence unable to escape alienation merely by changing personal outlook and attitude. The project of freeing the individual, as a social animal, must thus be tackled with other people.

Idealism and materialism

If radical social change is possible, how will it come; by first changing *material circumstances*, or most people's *ideas*, or both simultaneously? Where should the emphasis lie in the strategies of radical groups? Does what we think about nature condition what we do to it (White 1967), or does what we do to nature condition what we think about it (Thomas 1983, 23–5)?

An extreme idealist might claim that the world can be changed by thinking about it. If people decide, for instance, that it is a good idea to start behaving cooperatively, non-aggressively and benignly towards nature, then they can do so. If you want to change society in these directions, then you need to change attitudes and values, particularly those in the minds of people who run the institutions where we learn our values and ideologies – media and education, for instance. Thus Goldsmith (BBC 1987a), typically for greens, considers that action will change following changes in consciousness, as night follows day:

> I honestly believe that if people knew the truth about the pollution caused by nuclear power stations and the dangers of pesticide residues in their food they would not tolerate either the nuclear or the chemical industries.

An extreme materialist would argue along opposite lines. In particular, the economic organisation of society leads to particular social and economic relations between the people engaged in producing things. These in turn determine most people's ideas. Thus in days of slavery the beneficiaries – the slave owners – thought it obvious that they were more noble than their slaves; people who do well in a particular economic situation generally come to see it as being judicious and natural. So if people compete with each other (for jobs, resources, markets) and exploit nature (because this is inherent in the economic system) then these competitive, exploitative relationships will incline most people to *believe* that competition or nature exploitation are good, or common sense or 'natural', hence unavoidable. Only under different material circumstances will ideas radically incongruous with the current material basis of society become widely accepted, as distinct from being just 'countercultural' minority opinion.

Idealism, says Peet (1991, 51–2), was feudalism's finest intellectual achievement:

> Hegel's idealism connects individual consciousness with a collective and transcendent World Spirit. Movements of Spirit precede human thought and material events, in some way causing them. God 'wishes' an event to happen and moves in mysterious ways to effect it. . . . History is the evolution of an ever more perfect World Spirit.

And history today, as taught and interpreted in bourgeois cultures, is often

13

presented as parades of, and conflicts between, ideas – usually as articulated by 'great men'.

Marx and Engels (as spelled out in Chapter 3.2) stood this approach on its head:

> In direct contrast to German philosophy which descends from heaven to earth, here we ascend from earth to heaven . . . we do not set out from what men say, imagine, conceive; nor from men as narrated, thought of, imagined, conceived, in order to arrive at men in the flesh. We set out from real, active men, and on the basis of their real life-process we demonstrate the development of the ideological reflexes and echoes of this life process.
>
> (Marx and Engels 1981, cited in Peet 1991, 56)

Marxian materialism, seen thus, would appear to be 'extreme', though there are those (e.g. Cuff and Payne 1981, 58) who see it as a compromise between the two extremes – arguing that new ideas and consciousness *can* change the world provided that people act on them. But the extent that people *will* act on them will be conditioned by how much they are compatible with what people already are *doing* (and thus how much the new ideas will be seen as an acceptable extension of 'common sense').

This debate – essentially about strategy – has dogged much of red–green politics. Atkinson (1991) attempts to resolve it by a further movement away from materialism (influenced perhaps by the neo-Marxist 'Frankfurt School' which distanced itself from crude or dogmatic materialism). He describes (p. 6) the capitalist economic system as a set of cultural *attitudes*, believing that humanity's coming to the 'verge of self destruction can be traced rather directly to the radical separation of the objective and the subjective' (p. 45) – i.e. the development of (Cartesian) *ideas* during the Enlightenment. You cannot over-rely on materialist explanations of social change, he says, because 'there is no material influence on life that is not mediated by ideological structures' (p. 107). And ideas do not just reflect people's material interests; they form independently in response to our aesthetic preferences as well in response to our material circumstances, because people innately search for symmetry, coherence, harmony and order in their lives. Thus, historically, countless people have stuck to their *ideas* even though to do so damaged their material interests (p. 72). So Atkinson concludes that there is a 'dialectical' interplay between actions and ideas. What we do is influenced by ideas, social structure and relations, nature, aesthetic desires, and a sense of anticipation about the future (p. 59), and none of these is more important that the others.

However, in the end Atkinson perhaps comes down on the side of idealism, as greens are wont to do (pp. 113–14), seeing religious ideas as the crucial determinant in the formation of individualism, capitalist accumulation and exploitation of nature, after Max Weber. The Puritan work ethic conditioned people continuously *to produce* material things from earth's resources, but the simultaneous preaching of ascetic (severely abstinent) values restrained them from *consumption*. Hence came a mentality valuing *accumulation* as the ideal response to God's injunctions. This is, says Atkinson, a form of masochism, and the obverse side of it is sadism towards nature. Ecological disruption, in this view, becomes a psychosis giving rise to economic behaviour, rather than the other way round; an interpretation which has travelled so far from Marx as to be un-Marxist.

Collective or individual action?

Radical social change, achieved by confronting people's ideas or their economic organisation, means also confronting the political power of those who benefit from present arrangements. This power is so formidable that it might only be confronted by people acting *en masse* in conventional political ways, ranging from parliamentary politics to extra-parliamentary pressure group action or, more likely, revolution – withdrawing labour and/or seizing the instruments of power. All these routes favour *collective* approaches, by contrast with approaches which see all political change starting with the *individual*. According to the latter perspective, it is no good expending energy to get the masses to take political power if you yourself have not changed the way you think and live. This is because 'the personal is political' – a favourite green and feminist adage which means that all our thoughts and actions as individuals (e.g. in choosing the food we eat) have political ramifications. In a way this could be regarded as a collectivist view, because it emphasises how individuals are part of wider society. Yet in practice this implication of the adage is usually neglected in favour of the implicit suggestion that it is the individual self that has the *pivotal* role in social change. The individualist approach mistrusts mass revolution, arguing that it usually involves violence and oppression, the very things that revolution probably intended to conquer in the first place (though in the late 1980s, revolutionary changes in Eastern Europe entailed little violence). And it mistrusts party politics, arguing that the search for political power irrevocably corrupts politicians, and that political parties always have to compromise their ideals. Individualism places faith, instead, in a continuous process of individuals changing their values and lifestyles, which should then produce a new aggregate society. This concept rests on an essentially liberal view of society (see below).

In Britain, collective action for social change is most readily associated with the trades union and labour movement. But it could also imply the kind of local community politics which are effective on the European mainland, and are strongly advocated by the Green Party (Wall 1990).

However, collectivism is not fashionable in today's political climate. It is associated, says Griffiths (1990) with the establishment of the regulatory state in Britain in the second half of the nineteenth century, when *laissez-faire* was not regarded as a principle of sound legislation and government intervention was seen as beneficial – even when it limited individual choice or liberty. But today, we have 'problems more serious even than those of the mid-nineteenth century', social and ecological, and although their scale is so great that

> only the authority and resources of governments can begin to solve them
> . . . it is our tragic misfortune that the crisis has occurred when, under the prevailing political and economic philosophy, public and collective action is denigrated. . . . The present government is . . . wholly committed to this disastrous pursuit of self-interest.

Therefore the government welcomes as solutions deregulation, privatisation and capitalist adventurism, which actually create the problems in the first place.

Society: *gemeinschaft* or *gesellschaft*?

This question, of whether individual or collective social change strategies are best, relates to a more fundamental debate about what concepts of individual, society and community actually mean and imply – a debate which helps to define traditional political ideologies. It can be approached through sociologist Ferdinand Tonnies's (1887) distinction between the ideal types *gemeinschaft* and *gesellschaft*.

Gemeinschaft describes a social relationship founded on 'solidarity between individuals based on affection, kinship or membership of a community' (Bullock and Stallybrass 1977, 256). People have a sense of community which amounts to more than just the sum of the individual identities in it, and they explicitly or implicitly believe in their society as organic and based on unalienated face-to-face relationships. Conservatives and socialists share, at root, this ideal type. But the former go on to define it in ways that socialists do not approve of: involving hierarchy and status inequality as binding forces in the organic society, and harking back to feudalism.

Liberals, however, embrace *gesellschaft* social relations, involving 'division of labour and contracts between isolated individuals consulting their own self-interest' (Bullock and Stallybrass). Society, then, is atomised (cf. the Newtonian view of nature as composed fundamentally of individual atoms; a view which rose alongside liberal philosophy) and its totality amounts only to the sum of the individuals in it. Relationships are based on individual interests and rights, each person having equal rights to property, for instance. Maximum social good is thought to flow from individuals all seeking to maximise their own gain, after the 'invisible hand' theory of Adam Smith.

Kamenka (1982a, 8–24) describes how socialist community mores imply, after Rousseau, that the 'general will' is qualitatively different from the sum of individual wills, and the latter may have to be subordinate to the former. The general will is an expression of humanity's social, communal *nature*. To be fully human is to live with others and be concerned for them as one is for

16

oneself. Therefore to be separated from this communal aspect of self, through rampant individualism, is to be alienated. This theme runs strongly through Marx's and Morris's works.

In this socialist 'total community', property is social rather than private, labour has dignity, humans are equal, and austerity, modesty and devotion to the public good are virtuous. 'Pure' socialism therefore argues for a cooperative, unhierarchical and secular *gemeinschaft*, which Marx called *gemeinwesen* (ultimate communism). And even today's social democracy, which is far from its socialist roots, defines the political agenda partly on this basis of collectivism and public good:

> The reality is that, despite the atomisation of industrial societies – the breakdown of traditional communities, the extension of labour mobility, household self-sufficiency – the modern era demands more, not less, collective decision making and cooperative action, whether to protect the environment . . . or combat global poverty.
>
> (Blackstone, Cornford, Hewitt and Miliband 1992).

By contrast, the radical right's *gemeinschaft* revolves around the notion of 'natural' laws binding people in an organic (slow changing) unity, and binding people to nature. The latter comes out particularly in the nationalist conception of intimate links between people and 'their' soil, landscape and folk traditions (Mosse 1982). Such links are romanticised in visions of pre-industrial medieval and 'traditional' societies. The idea of the community therefore grows out of the people; their locality and shared material existence. The source of authority is the general will, but since it is a natural hierarchy, that will is expressed through leaders. Both the bioregionalism of deep ecology (Chapter 4.5) and the utopian environmentalism of Goldsmith (1988) stress the need to re-establish such values of small-scale pre-industrial traditional societies. They go beyond rational expression, being articulated in nature mysticism, creative art, folk legend and paganism.

Liberals bow to the collective if they get something out of it for themselves. But they see human nature as *autonomous* – having standards and principles which are unique to the self (Benn 1982). Hence they do not regard the collectivity as something which soars above the sum total of selves, and will not accept collective mores without subjecting them to rational, critical and suspicious scrutiny. In Margaret Thatcher's infamous aphorism: 'There is no such thing as society. There are individual men and women and there are families' (cited in the *Observer*, 27 December 1987).

So any concession to 'society' in the form of cooperation with its laws, morals, or economic and social arrangements, is predicated on strict reciprocity and mutuality. Liberals talk much of 'contracts', social and otherwise. People monitor their own behaviour towards others, and adjust it conditionally, depending on how others treat them. *Gesellschaft* is the minimalist conception of community which most people share in Western liberal, capitalist nations.

17

Its project – of securing maximum freedom for the individual – is very much an Enlightenment ideal. Atkinson (1991, 153–8) traces it to the Reformation, which told people not to follow blindly the Church, but the dictates of their individual conscience. The individual grew to be treated as the absolute origin of knowledge and action. Individualism was also enshrined in the Calvinist doctrine of self-responsibility. The *ultimate* source of all this was pre-Enlightenment: in the Judaic doctrine of the human soul as the focus of ultimate value, meaning and salvation – a doctrine which infuses Cartesian rationalism and Anglo-Saxon morality (Atkinson 1991, 159).

It should be emphasised that socialism shares this Enlightenment individualist ideal. The difference between socialism's interpretation of it and that of liberalism is that socialism sees maximum *individual* freedom coming through fulfilment of the *collective* side of the individual's nature.

The call to a greater sense of 'community' suffuses green literature also. But it is not easy to decide just which notion of community is most 'green'. As has been suggested, both conservative and socialist *gemeinschaft* are relevant (e.g. Goldsmith or Ryle, both 1988). So, too, is *gesellschaft*. There seems to be no consistent green ideal type.

However, it may be of more than passing interest that the faults and vices of over-individualism, as detailed by writers on the left, often seem to be prominent among the faults and vices of the green movement (Chapter 3.9). There is the fault of 'apoliticism', for instance, where collectivism is denied by spreading the belief that the problems of capitalism cannot be resolved through collective political action but through individual reform: Hence 'There are no social evils: there can only be evil people' (Seabrook 1990).

Related to this is the problem of guilt inculcation. As Seabrook again says, the 'noble' Western project of cherishing the individual leads to the prevalent ideology that individuals are primarily *responsible* for their own wellbeing. Therefore if things go wrong (e.g. with their environment), responsibility and repair must come through individual (lifestyle) reform. Society and social determinants are eclipsed, while collective political action is avoided and derided. Then, when individual lifestyle reform does not radically improve things, individuals feel guilty. Obviously they did not try hard enough. Guilt culture, thinks Atkinson (1991, 159–61), is specifically a facet of individualism. The worth of actions is judged by the actor, so ineffective actions are the actor's fault, requiring repent. However, communal cultures generate not guilt, but shame in the face of the rest of the group, and the need to redress. In guilt there is a counterproductive, existential separation of the 'I' from the rest of the community.

Then there is the narcissism which results from the nineteenth- and twentieth-century trends away from an 'outer directed' to an 'inner directed' society: i.e. a society which values 'public' (collective) life over one which centres on the private (individual). Sennett (1978) describes narcissism as a *self-absorption* – an obsession with what other people and events mean to *me*

18

rather than to *us* or to society, community or group. Ideas of love and commitment have been redefined during the social changes which accompanied the rise of liberal capitalism. They now equate gratification with meeting *self*-needs, and commitment to wider political reforms as realisable primarily through commitment to *self*-discovery – the vital first and most important step.

Self-knowledge and self-disclosure to others have, indeed, become hallmarks of the 'green way'. New Ageism is their highest development, but nearly all ecologism displays some of these obsessions.

Sennett considers that nowadays self-disclosure is part of a 'destructive *gemeinschaft*' because it is seen as a moral good in itself, making a tissue which binds people, irrespective of the social conditions which form the context of such disclosures. It creates a sort of community – a collective identity but one which is not based on collective *action* for social change as the binding force. What people have in common is their 'openness' about themselves and how they feel, and their search for *self*-understanding and fulfilment. They do not necessarily have in common shared action for shared political and social goals – indeed they often shun such action.

Consensus or conflict: pluralism, elitism or Marxism?

Rejecting collective political action and embracing individualist lifestylism often goes hand in hand with rejecting a conflict model of social change. Proponents of conflict models will argue about the inevitability of conflict in any radical social change process. Groups which want to change society will have to face up to the fact that there will be conflict between those who have power and do not want to give it up, and those who seek power. Hence there may be conflict between 'ruling class' and 'employee class', or between men and women, or between different race/ethnic groups, or geographical core regions and peripheries, and so on. One important conflict model is that of Marxism, which argues that although society may be structured into classes or groups in various ways, *two* classes particularly are significant in the change from capitalism to socialism. Despite the complications of the rise of the middle classes and widespread share ownership, it may be still broadly possible in advanced capitalism to think in terms of those who effectively own and control the means of production (including natural resources), distribution and exchange, and those who do not, and have only their labour to sell. This conflict perspective sees social change arising from the inherent, latent struggle between these groups. And since this is the struggle that has been the main concern of socialists and the labour movement, it would follow to Marxists that new energies for social change – that come through green concerns, for instance – should be directed through these traditional channels. Therefore, anyone who wants to change society should show a consciousness of how their new concerns relate to the class struggle: the 'old' politics of poverty and wealth, left and right.

Many regard this approach as simplistic and/or denying the idea that we live in a democratic, *pluralist* society. This is composed of a plurality of groups, all related in a system, and when one group is particularly alienated or disadvantaged the system will adjust: not through revolutionary conflict but through appeal to the law, or through government responding to pressure group protest, or firms responding to consumer pressure and so on – to lessen that group's grievances. Thus a new consensus is reached and the system remains stable, though changing and evolving. It will be the task of any new interest group, articulating new concerns, to enter and use this process by pressure group politics, applying rational argument and 'reasonableness' in their lobbying. This will challenge and change the old consensus, yet the broad parameters and structures of social policy and decision making will remain in place. Like any natural system, the social system in this model remains robust and stable by accommodation and adjustment, not by being forcibly impelled, via positive feedbacks, over new thresholds. Hence the judicial, parliamentary and bureaucratic system in a Western 'democracy' like Britain is ostensibly based on the idea that when two sides dispute something a satisfactory resolution will not necessarily involve 'natural justice' but an outcome where each side gets *something* of what it wants.

Pluralism, then, believes that 'democracies' are indeed democratic. In them all citizens have the right to seek, and the opportunity of seeking, access to the political process in pursuing their own preferences. Disputes (say, about environmental matters) are settled within a planning system which has a consensus of support. So, by implication, do the decisions which are reached. And, as Bullock and Stallybrass (1977) put it, the concept of pluralism is

> frequently used to denote any situation in which no particular political, ideological, cultural or ethnic group is dominant. Such a situation normally involves competition between rival elites or interest groups, and the plural society in which it arises is often contrasted with a society dominated by a single elite where such competition is not free to develop.

In this latter, *elitist*, view, society is indeed composed of competing interest groups, but the process of accommodation to each group is biased, towards *particular* groups who have 'unfair' (from a pluralist perspective) advantages. Thus the resources of money, articulacy, education and time which the environmental movement possesses may give it an overwhelming advantage over the interests of those in lower socio-economic groups when it comes to decisions about where some environmentally damaging project may be placed (see, for example, case studies in Kimber and Richardson 1974).

A *Marxist* view takes this elitist analysis one stage further, accommodating it to a conflict model based on material economic interests. Hence the fact that a particular group *is* an elite, able to manipulate the system to its own advantage, is thought to be based specifically on that group's *economic* power. For it follows in Marxist analysis that the division of labour under capitalism inhibits upward mobility between economic classes, and maintains a non-egalitarian society where the ruling capital-owning classes are more enabled than others to realise their own interests. So structural constraints operate in favour of the ruling class and against oppressed pressure groups – the techniques of planning, for example, reflect and reinforce the social order and world view of capitalism. Thus it is naive for environmental protest groups to appeal to supposedly neutral authorities who are set up ostensibly to balance and reconcile conflicting interests. For these authorities, such as planners or members of Parliament, *cannot* act as environmental managers in a way which is free from the constraints of a social–economic structure which is designed to further the interests of capital.

Pluralism implies that capitalism is broadly democratic, elitism that it is not but might be, and Marxism that it is not and cannot be. There is much evidence against the pluralist perspective on environmental issues. Hamer (1987), for instance, shows how the road transport lobby in Britain controls Parliamentary decision making on transport, while Blowers and Lowry (1987) demonstrate that the Anglo-American nuclear industry shapes scientific research and central government decisions on nuclear power and waste. Blowers (1984), in analysing the history of decision making about the siting and scale of Bedfordshire's brickworks, finds that elements of pluralist, elitist and Marxist models all apply to that case study at different times.

Structuralism

The above suggests an obvious, but sometimes neglected point. We must think clearly about how society works, because our theory on this will – or should – determine the appropriate *strategy* for changing society. This is further illustrated by the question of *structuralism*. Structuralism asks whether what we see in the way of social events and individual and group behaviour is to be interpreted in terms of deeper and less apparent underlying structures in the human mind and/or in society. In Chomsky's terminology, are we to regard surface structures as conditioned by underlying ones? Or is what we see simply all that there is?

Our answer to this question will obviously determine whether we work on what we see, or on what we think underlies what we see. And if we do adopt structuralist methods, then we have to decide what are the significant underlying structures – are they cultural or economic or consisting of the basic characteristics of the mind? So structuralism is a perspective preoccupied not simply with structures, but such structures as can be held to *underlie* and *generate* the phenomena that we see (Bullock and Stallybrass 1977, 607). It is, expressed thus, a deterministic and also reductionist concept. This is a danger,

21

for structuralism can lead us to reduce the world to a matter of what we perceive to be its deep underlying principles, denying any independent value or meaning which the surface level might have. The opposite danger might lie in postmodernist perspectives (see Chapter 2.4) which deny that observed phenomena reflect any deeper underlying principle.

In a broad sense, any theory is 'structuralist' if it posits that deep, unobservable, only subconsciously apprehended realities give rise to observable realities. This describes Marxist theory, which tries to discover the causes behind social events, as well as social relationships hidden in apparent 'objects' such as commodities (Chapter 3.3). These causes and relations are seen, in Marxism, largely in materialist terms. The *material*, economic class structure is thought to strongly influence the roles which groups and institutions play. And different economic modes of production (feudalism, capitalism, socialism, for instance) each have different overall social goals that relate strongly to the economic (hence political) aspirations of their dominant classes. Some structural Marxists (e.g. Althusser) make this model quite economistic (i.e. reducible entirely to economic factors) – that is, it conceives of a 'superstructure' of social beliefs, ideas, relations, institutions, practices and rituals which is quite rigidly *determined* by the underlying economic 'base'. For other Marxists (Peet 1991, 176) the relationship is more subtle and dialectical. We are not all drones, behaving strictly according to our economic class imperatives, and individual volition and consciousness is allowed in social theory (see Chapter 3.2). But structural Marxism never allows the substantial role of material, economic forces and structures to be forgotten either.

However, humanist Marxists, like Hogsbawm, Eagleton and Bordeiu, distance themselves considerably from economism and the theory that material interests have primacy in social explanation. They fear that structural Marxism dismisses human subjects and the possibility that they can consciously shape events, so they argue against the determinism of Althusser's assertion that history is simply a 'process without a subject'.

Atkinson (1991), from an ecological perspective, is more interested in the structuralism pioneered by Lévi-Strauss and inspired by Saussure's theories of linguistics. Lévi-Strauss was concerned with relating behaviour and institutions to basic characteristics of the mind, reflecting how mind imposes structures onto reality:

> the unconscious activity of the mind consists in imposing forms on content, and if these forms are fundamentally the same for all minds – ancient and modern, primitive and civilised (as the study of the symbolic function, expressed in language, so strikingly indicates) – it is necessary and sufficient to grasp the unconscious structure underlying each institution and each custom, in order to obtain a principle of interpretation valid for other institutions and other customs. . . .
>
> (Lévi-Strauss, *Structural Anthropology*, cited in Atkinson p. 73)

Thus, Lévi-Strauss argues that the apparent meaning and order of the natural world is not innate. It is humans who impose that order, through a mental capacity to classify. And, since all humans (he argues) have the same kind of brain, including those who live in 'advanced', 'civilised' societies and those who do not, then the mental organisation of structure is *universal*: everywhere is a tendency, for instance, to create binary opposites and see one side (hot, clean) in terms of the other (cold, dirty). Thus the symbolic meanings which Lévi-Strauss saw in human social behaviour (both on the surface of everyday appearance and in the deeper underlying structures) may be specific to given societies – but they are also *reiterated* through all other cultural subsystems.

His object was to understand the workings of the human mind, by analysing different cultures and how they attribute symbolic meaning. It was not to learn about the social organisation of any particular society, so it was 'ahistoric' (Chapter 3.2). Herein lies a principal division between structuralist approaches. On the one hand are versions that regard structures whose limiting logic resides in the human mind. On the other hand are those, like structural Marxists, which locate their structures within human society and particular economic, political and social arrangements (all subsumed under the term 'modes of production') which vary through space and human history, so that they cannot be described as 'universal'.

Development theories

If concern to change society necessarily entails a theoretical perspective on what causes social events and behaviour, it also requires a preference for particular social development models. Does the desired social change support an existing model, or does it constitute a new one? Peet (1991) explores the differences between existing development models with clarity.

Ecocentrism rejects strongly the prevailing model, in which 'development', for instance in the third world, is equated simply with *modernisation*. It observes that a dependent development model more accurately describes what is actually happening within and between nations. It prefers, however, a 'sustainable development' model. This, in its radical form (Engel and Engel 1990), combines independent development with bioregionalism. The latter incorporates elements of environmental determinism.

Environmental determinism (see above) holds that differences in how societies develop come down to differences in natural environment. Thus, in social Spencerism/Darwinism, some societies have greater 'natural' advantages in the struggle for survival. 'Rich' natural environments favour 'super-organic' evolution, to great population and political size and armed strength, economic specialisation and division of labour.

Obviously, such a theory gave pseudo-scientific legitimation to the fact that during the nineteenth century European people (claiming to be highly civilised due to climatic and other natural factors) extended their share of direct control over world space from 35 to 85 per cent, 'naturally' eliminating or subordinating 'less civilised' ethnic groups in the process (Peet 1991, 18). Environmental determinism also propped up the Third Reich's *lebensraum* ideology.

Though the bioregional development model (Chapter 4.5) specifically rejects theories of environmentally or biologically derived racial/national superiority and aggrandisement, nonetheless it clearly resonates with environmental determinism. For it advocates living within natural limits to growth, and largely according to such provisions of the natural environment as are found within bioregions. Furthermore, like environmental determinism and social Darwinism, bioregionalism stresses analogies between social and biological processes.

So, too, does structural functionalism. This regards a culture or society as an entity, all the parts of which function to maintain one another and the totality. If one part is disrupted this provokes readjustments among the others. Clearly this is a closed systems view of society, and it relates to the political concept of pluralism. Hence, what each group does is understood by reference to the functions which it has in relation to the other groups forming the whole structure. And activities and events can be understood as not simply random, but in their relation to the groups performing or advocating them, and to the whole structure. Ritual and ceremony, for instance, are supposed, according to Durkheim and Malinowski, to act to reinforce shared beliefs and practices – and the more that people *do* share beliefs and practices, the more stable is the society. The church or educational institutions can also have important stabilising functions as propagators of shared beliefs. Post-war structural functionalist sociology was heavily influenced by Talcott Parsons (Peet 1991, 22–8). He saw any society as analysable in terms of four functions. Its *culture* particularly functioned to maintain the patterns by which it is controlled (e.g. a social hierarchy). Its *social systems* integrated people and groups playing different roles, to keep the whole together. Its *personality systems* controlled goal attainment, so that individuals fulfilled roles (occupied niches), helping the whole society to attain its goals. And *behavioural systems* enabled society and its components to adapt to their environment.

Hierarchy figured strongly in all this – the relations between the systems being organised through a hierarchy of controls: so 'high order' (high in information, low in energy) systems regulated 'low order' ones (high energy, low information). The whole purpose was to maintain systems stability within an overall pattern of evolution: a dynamic equilibrium.

The appeal of these ideas to conservative ecocentrics is obvious, rooted as they are in biological systems theory and economic input–output models. Applying them, Goldsmith (1978) held up India's caste system as a model for attaining a socially and ecologically sound society (see Chapter 2.3). Such theories do not merely comment on the system of social order: they see it as desirable in the process of social survival and development. Structural-functionalism, then, is open to the charge of being politically reactionary. It argues against change because it describes systems as having needs, sees human subjects and what they do as the outcome of the structure of the system and so finds events predetermined.

These criticisms are similar to those which have been levelled against structural Marxism (see above). Peet points out (1991, 175–6) that the two are often equated, Marxism being seen as 'economic functionalism of a teleological [having design and purpose] variety' – that is, it says that groups behave and societies are shaped according to economic position and vested economic interests. But he insists that this is not accurate, for Marxism's purpose is radical revolutionary *change*, and all but the crudest forms play down determinism, encompassing

> individual volition, system-changing activities and the sense of imminent transformation. . . . Structural functionalism and Marxism are on opposite sides of the political fence, the one being a leading instance of legitimation theory [theory which justifies the status quo], the other a leading revolutionary theory.

Structural functionalism's offspring is modernisation theory. In the former, any change comes about gradually, through structural differentiation to new levels of adaption. This leads to eventual breakthrough by some societies to new levels of adaptive capacity. Modernisation theory holds that the more structurally specialised and differentiated a society is, the more *modern* it is. Modernisation involves technological sophistication, urbanisation, the spread of markets, democracy and education, social and economic mobility and the weakening of traditional elites, collectivities and kinships. Individualism and self-advancement attitudes prevail, but are tied in with the notion of overall social progress – the two being related by an 'invisible hand' after Adam Smith.

This theory proposes Rostow's (1960) stages of economic growth as an alternative to the (also progressivist) Marxist theory of history. It describes how 'traditional' societies (primitive technologies, spiritual attitudes to nature), 'develop' to 'pre-conditions for economic take off' (as experienced in seventeenth- and eighteenth-century Western Europe). 'Take off' follows, where new industries and entrepreneurial classes emerge. In 'maturity' steady economic growth outstrips population growth, then a 'final stage of high mass consumption' allows the emergence of social welfare.

Criticisms of this model are many. They range from objections to its teleological overtones, that see all changes fulfilling some grand design or systemic need, to rejection of its ahistorical perspective – seeing modernisation as a universal tendency and pattern. But, most obviously, the model is Eurocentric and imperialist, and it legitimises capitalism as 'progress' *per se*. Third world societies, particularly, and all communities (actual or potential) which base their economic and social relations on localism and kinship are, by definition, 'backward'. Their 'development' must come by opening up borders to the influence of Western economic interests, via arrangements like GATT.

As I have said, ecologism explicitly rejects modernism theory. But Peet (1991, 41) interestingly implies that the New Age tendency in ecologism (see Pepper 1991) is in fact an example of restatement of modernisation theory, laying stress not on developing new social behaviour and practices but revolution in and modernisation of *consciousness*. As such this still emphasises the cultural visions of an elite: middle-class preoccupations are what defines that which is deemed modern, of the 'new age'.

Greens, anarchists and the neo-Marxists of the new left unanimously see that 'modernisation' is really *dependent development* in disguise. That is, 'underdeveloped' nations increasingly depend on political and economic relations with 'developed' nations for their livelihood, and vice versa. And this does not produce a set of developed nations helping underdeveloped ones to catch up – to modernise. For, just as poverty is a necessary feature of capitalism (Seabrook 1989, 1990), so too, in capitalist world 'development', underdeveloped nations are essential counterparts of the existence of developed ones. In other words, underdevelopment *results from* and is a vital feature of capitalism, which is a way of transferring wealth from peripheries to core areas, within and between nations. World-wide 'free' trade helps this appropriation of surplus, accentuating regional disparities.

These critics therefore propose *independent development* models, including sustainable development and bioregionalism, where regions encourage each area to go their own economic way, cutting down trade and specialisation. The important neoclassical economic principle of complementarity (countries specialise in what they do best and exchange their products with other countries, who have specialised in what they do best) would be abandoned, so making nations far less beholden to each other and to world market prices. This model has all sorts of political as well as economic ramifications, most of which greens and anarchists welcome (see, for instance, Schumacher 1973, Sale 1985).

But for most Marxists both dependent and independent models are too simplistic. The latter are also politically unacceptable. Dependent development models are simplistic because they neglect the observable fact that it is possible for 'peripheries' to show higher growth rates than cores, as is happening today (compare Western Europe with some Far East countries). And independent development glosses over the factor of class exploitation, which is a marked feature of Asian growth and 'modernisation' today and produces sophisticated electrical consumer goods for Western markets through sweated labour, for instance. This is politically bad, for neither 'modernisation' nor independent development are acceptable if exploitative productive relations are allowed to persist. Nothing short of a development replacing capitalism by socialism is therefore allowable.

To a structural Marxist, like Peet (1991, 73–7), the true development picture is one of the *articulation*, or joining together, of different modes of production in a world system. Hence the picture is multilinear, not unilinear: there are several, often conflicting, developmental tendencies. But, in most, ruling elites extract surpluses. This analysis provides the basis for Peet's exposition of desired development along the lines of Marx's historical materialist model (Chapter 3.2 and 3.7), incorporating, however, new emphases on ecological and feminist concerns.

Poverty, egalitarianism and market intervention

These are central concerns in the 'old politics' which greens often claim to reject. Yet they were always at issue in modern environmentalism too, where they have lately occupied centre stage. This is not so much in response to the pleas of those socialists who argue that the labour movement's crusade against poverty is and always was essentially an *environmental* crusade (Weston 1986). Rather, it reflects the growing tendency to articulate third world concerns as central to those of Western ecologism.

The Brundtland Report (UN 1987), particularly, publicised the third world '. . . vicious circle of poverty leading to environmental degradation, which in turn leads to even greater poverty' (p. 31). In the 1960s, 18.5m people were affected by the environmental disaster of drought, and 5.2m by floods. In the 1970s this had risen to 24.4m and 15.4m respectively, and Brundtland believed that the 1980s would see the trend accelerating. It said:

> Such disasters claim most of their victims among the impoverished in poor nations, where subsistence farmers *must* make their land more liable to drought and flood by clearing marginal areas, and where the poor make themselves vulnerable to all disasters by living on steep slopes and unprotected shores – the only land left for their shanties.
>
> (UN 1987, 30; emphasis added)

Brundtland identifies unequal land distribution, growing demand for commercial (cash crop) rather than subsistence use of land and rapid population rise as the reasons why subsistence farmers have been pushed onto marginal land and shifting cultivators have less land and time for their rotations.

If poverty, then, is both cause and result of environmental problems, its elimination must become the green imperative. For 'There are more hungry people in the world today than ever before and their numbers are growing' (ibid., 29). In 1980, 340m people in developing countries were not getting enough calories to prevent stunted growth and serious health risks. In 1992, a statement from the Royal Society of London and the US National Academy of Sciences identified one billion people in absolute poverty, mostly in the third world, while developed countries have 85 per cent of the world's GNP, and account for most of the world's mineral and fossil-fuel consumption, but have only 23 per cent of the population (*Guardian*, 27 February 1992; see also Table 1.2).

But the greens have developed no new response to poverty, except to

27

Table 1.2 The balance of progress and deprivation

Environment Guardian, 24 May 1991
From *The Human Development Report*, UN Development Programme, May 1991 and the UN Population Fund Report, May 1991.

The *Guardian*'s commentary, by John Vidal, concludes: 'Thirty years ago agricultural and industrial development was seen as the global problem by world bodies, then it was population, and more latterly the environment has been billed as the great question mark for humanity. What both UN reports illustrate – and perhaps this is the greatest progress made lately in developmental thinking – is that all three are linked and all lead inevitably to desperate poverty. Meanwhile the inequalities only increase.'

PROGRESS DEVELOPING COUNTRIES	DEPRIVATION	PROGRESS INDUSTRIAL COUNTRIES	DEPRIVATION
LIFE EXPECTANCY Life expectancy increased by one third between 1960 and 1990. Now 63 years.	Ten million older and 14 million young children die each year.	**LIFE EXPECTANCY** Expectancy 75 years. Virtually all births attended by health workers.	
HEALTH Proportion of people with access to health services has risen to 63 per cent. Proportion of people in rural areas with access to adequate sanitation has doubled in ten years. Eighty-eight per cent of urban dwellers have access to health care and 81 per cent have access to safe water.	A total of 1.5 billion lack basic health care or safe water. Over 2 billion without safe sanitation. Only 44 per cent of the developing world has access to basic health care.	**HEALTH** Two-thirds of the population covered by public heath insurance. Only 8.3 per cent GNP spent on health care.	Adults smoke average 1,800 cigarettes and consume four litres pure alcohol a year. More than 50 per cent likely to die of circulatory and respiratory diseases linked with sedentary lifestyles and nutrition.
EDUCATION Adult literacy in developing countries rose from 46 to 60 per cent, 1970–85.	One billion adults illiterate and 300 million children not in school.	**EDUCATION** Six per cent GNP spent on education.	Almost one in four lack secondary education.

INCOME

Income per head grew annually in 1980s by 4–9 per cent in east Asia. More than one in four people live in countries with growth rates above 5 per cent.

More than 1 billion live in absolute poverty. Income per head has declined in the last decade in Latin America and sub-Saharan Africa.

INCOME

GNP per capita increased 1976–88 almost 300 per cent. These countries produce 85 per cent global wealth. Social welfare expenditures now 11 per cent GNP.

The wealthiest 20 per cent receive almost seven times as much income as poorest 20 per cent. About 100 million people live below the poverty line.

POPULATION

Under-five mortality rates halved over last 30 years. Immunisation coverage increased dramatically in 1980s.

One-fifth of the world's population still goes hungry every day. About 14 million children die before they are five years old, and 180 million suffer chronic, severe malnutrition.

POPULATION

Growth rate around 0.5 per cent. Almost everyone enjoys access to safe water and sanitation.

Up to 50 per cent dependency ratio. Total of 433 persons in 100,000 are seriously injured in road accidents every year.

---

eulogise non-material aspects of wealth. A few even express a traditional conservative view, of a hierarchical, unequal society in which, however, the better-off have moral responsibility for the poor (and for nature). Many more, however, take the liberal meritocratic view, that everyone should have equal opportunities, yet this will produce differences in wealth because of inherent differences in people's abilities – however, the differences can be offset by devices like a basic income scheme. Other, communalist, greens aspire to the socialist position, that notwithstanding inherent differences everyone should be valued and treated equally from the outset, gross inequalities not being allowed to develop – hence the maxim 'from each according to ability: to each according to need'.

Most Western democracies have liberal political parties whatever their label, and West European greens have publicly distanced themselves from socialism. Hence they mainly advocate or tolerate capitalism, differing largely on how free the market should be. It is the minority socialist or anarchist groups, like the Socialist Party of Great Britain (SPGB), who would do away with capitalism and a money economy, hence eliminating accumulation and inherent build-up of wealth differences. (Although some green local currency schemes have been devised with inbuilt mechanisms that penalise accumulation, like automatic devaluation or regular changes of currency.)

Right-wing liberals believe that minimal intervention in market economics will produce the maximum wealth for society as a whole, and this will trickle down to the poorest, thus fulfilling the moral imperative of eliminating poverty. Faith in this mechanism has been extended to moral and pragmatic imperatives towards the environment. Simon and Kahn (1984) have exhaustively asserted how free market economics could overcome pollution and resource shortages, while Britain's former environment minister, Michael Heseltine, in supporting a voluntarist position on industry and pollution, claims that 'Cleaning yesterday's environment, sustaining today's environment and protecting tomorrow's is a moral obligation. But morality can march along the same road as the capitalist system' (*Guardian*, 6 November 1991).

Most greens, however, from technocentrics (Pearce *et al.* 1989) to ecocentrics (Porritt 1984) (see Chapter 2), agree that the unfettered market cannot protect common wealth and resources, and they join welfare liberals and democratic socialists in advocating degrees of state intervention in the system.

The long-apparent fact that the trickle-down theory does not actually work continues to be in evidence, even among the rich Western nations. Thus, while Brundtland asserts (UN 1987, 29) that 'The number of people living in slums and shanty towns is rising, not falling' in the third world, government statistics demonstrate that the number of British families living on or below the official poverty line increased by more than half to 6.2m during the first eight years of the Thatcher government, which vigorously promoted 'free' market economics (Department of Employment New Earnings Survey for 1990). And the Central Statistical Office's *Economic Trends* showed that a trend towards greater wealth equality in Britain begun between the wars has now been reversed. The richest people's wealth share is now higher than it was in the 1980s, being no less than it was in 1976 (*Independent*, 3 January 1992). Specifically, 6.7 per cent of people possess 61 per cent of personally owned capital, 85 per cent of shares and 81 per cent of land (Buick 1992).

Trickle-down theorists might counter that, despite this, the poorer are better off than they were, say, a hundred years ago. Yet there has been an absolute increase in total global poverty with the spread of capitalism, while in the West it is the distribution, not the level, of incomes that matters most in determining people's health and life expectation as well as the degree of control over their lives (Donnison 1991). All of these are vital parameters of environmental quality.

Thus greens are enmeshed in a very old and problematic political issue. But they also bring a 'new' dimension of it to the political agenda: that of biological egalitarianism. The notion of a democracy among all creatures is very old philosophically, but deep greens, like Earth First! and animal rights activists, have attempted to make it a contemporary political matter. Biological egalitarianism challenges human centredness in economics and development, arguing instead for biocentrism. The lack of wide political appeal in this message (except perhaps over factory farming) is fundamental. For biocentrists have to overcome the attitude inherited from the Enlightenment, that nature should be used for human material benefit, plus the fact that human survival inevitably hinges on a certain amount of killing and exploitation of nature. Furthermore, it is impossible for humans not to be *anthro*pocentric: perceiving nature from the perspective of human consciousness. Hence the biological egalitarian's position has for realpolitik purposes, to become more anthropocentric – albeit a 'weak' anthropocentrism benign to nature rather than the avoidable 'strong' anthropocentrism that uses the non-human world merely as a means to an end (Dobson 1990, 60–6).

2

POLITICAL ECONOMY AND POLITICAL IDEOLOGY: WHERE GREENS, MARXISTS AND ANARCHISTS FIT IN

2.1 ECOCENTRISM AND TECHNOCENTRISM

Before considering how anarchism and Marxism may illuminate green thinking it is useful to extend the discussion of context which was begun in Chapter 1.2 by examining how these three relate to each other in the sense of political economy and traditional political ideologies. To try to do this immediately invites condemnation from those who see ecologism as a new and separate political ideology. But I think that Chapter 1 has demonstrated that while ecologism may *start* from different premises and concerns to those of traditional politics, it has to become involved in old political questions when it begins to say what we should do to attain ecological rectitude. Hence, ecologism can at least partly be analysed in terms of the classic questions posed by political economy, and be mapped against other ideologies. The exercise is instructive because, first, it helps to define what 'greens' we are and are not discussing in this book. Secondly, it suggests that there are, indeed, grounds for concern on the part of those green activists who believe that their political ideology is too eclectic or lacking in coherence:

> the role of pressure groups has always had inherent weaknesses: to concentrate on pushing the establishment in a certain direction fails to challenge their power head on. . . . Those in power have also welcomed the pressure groups with suspiciously open arms, seeing . . . a relatively cheap method of courting popularity. . . . Pressure group activity in a vacuum, without an ideological framework or long-term strategy for change, is all too prone to exploitation by the Establishment. . . .
>
> (Andrewes 1991)

If environmentalism is about ideologies and practices which flow from a concern for the environment, it is no exaggeration to say that most politically aware people in the West are to an extent environmentalists now. However, some are 'light' green; others are 'deeper' green. Since colours are relative to everyone, it is wise not to use such terms, but a classification like O'Riordan's

(1981), which is still very useful. He proposes a fundamental, but not mutually exclusive, division between 'technocentric' and 'ecocentric' perspectives.

Ecocentrism views humankind as part of a global ecosystem, and subject to ecological laws. These, and the demands of an ecologically-based morality, constrain human action, particularly through imposing limits to economic and population growth. There is also a strong sense of respect for nature in its *own* right, as well as for pragmatic 'systems' reasons.

This 'bioethic', which prioritises non-human nature or at least places it on a par with humanity, is, as Eckersley (1992) stresses, the key aspect of ecocentrism. It distinguishes ecocentrism from the anthropocentrism of other political ideologies, including socialism and anarchism.

Ecocentrics lack faith in modern large-scale technology and the technical and bureaucratic elites, and they abhor centralisation and materialism. If politically to the right they may emphasise the idea of limits, advocating compulsory restraints on human breeding, levels of resource consumption and access to nature's 'commons'. If to the left, their emphasis may be more on decentralised, democratic, small-scale communities using 'soft' technology and renewable energy, 'acting locally and thinking globally'.

The ecocentric position on technology is complex. On the whole it is not anti-technology, though it is 'Luddite', when one remembers that the Luddites did not protest against technology of itself but against its ownership and control in the hands of an elite. Ecocentrism advocates 'soft', 'intermediate' and 'appropriate' – that is, 'alternative' – technologies partly because they are considered more environmentally benign, but also because they are potentially 'democratic'. That is, they can be owned, understood, maintained and used by individuals and groups with little economic or political power, unlike high technology.

Hence ecocentrism is concerned with the Marxian idea that different technologies embody different forms of social relationship. Information technology, nuclear power and modern green revolution agricultural techniques are examples of technologies which were very much born of a society that is best described by the elitist or Marxist models. In it, these technologies facilitate economic, social and political relationships of hierarchy, domination and control (Albury and Schwartz 1982). Ecocentrics have emphatically condemned the last two technologies, but are more ambiguous about the former. There is a huge literature about the role of technology in a technology-dominated society – the relevance of this debate to ecocentric concerns is well described in Mumford (1934), Schumacher (1973) and Winner (1986).

Technocentrism recognises environmental problems but believes either unrestrainedly that our current form of society will always solve them and achieve unlimited growth (the 'cornucopian' view) or, more cautiously, that by careful economic and environmental management they can be negotiated (the 'accommodators'). In either case considerable faith is placed in the usefulness of classical science, technology, conventional economic reasoning (e.g. cost–

Table 2.1 European perspectives on environmental politics and resource management: contemporary trends in environmentalism

Ecocentrism		Technocentrism	
Gaianism	*Communalism*	*Accommodation*	*Intervention*
Faith in the rights of nature and of the essential need for co-evolution of human and natural ethics.	Faith in the cooperative capabilities of societies to establish self-reliant communities based on renewable resource use and appropriate technologies.	Faith in the adaptability of institutions and approaches to assessment and evaluation to accommodate to environmental demands.	Faith in the application of science, market forces, and managerial ingenuity.
'Green' supporters; radical philosophers.	Radical socialists; committed youth; radical–liberal politicians; intellectual environmentalists.	Middle-ranking executives; environmental scientists; white-collar trade unions; liberal–socialist politicans.	Business and finance managers; skilled workers; self-employed; right-wing politicians; career-focused youth.
0.1–3 per cent of various opinion surveys.	5–10 per cent of various opinion surveys.	55–70 per cent of various opinion surveys.	10–35 per cent of various opinion surveys.
Demand for redistribution of power towards a decentralised, federated economy with more emphasis on informal economic and social transactions and the pursuit of participatory justice.		Belief in the retention of the status quo in the existing structure of political power, but a demand for more responsiveness and accountability in political, regulatory, planning and educational institutions.	

Source: O'Riordan (1989).

benefit analysis), and the ability of their practitioners. There is little desire for genuine public participation in decision making, especially to the right of this ideology, or for debates about values. The technocentric's veil of optimism can be stripped away to reveal 'uncertainty, prevarication and tendency to error' (O'Riordan 1981). Technocentrics envisage no radical alteration of social, economic or political structures, although those on the left are gradualist reformers.

O'Riordan (1989) has elaborated this classification (see Table 2.1 and Figure 2.1). He points up how technocentrism is politically reformist by comparison with ecocentrism, which requires a radical redistribution of political power. He also describes technocentrism as 'manipulative': that is, regarding humanity's task as one of manipulation and transforming nature into a 'designed garden' to improve both nature and society. But there are differences between those

34

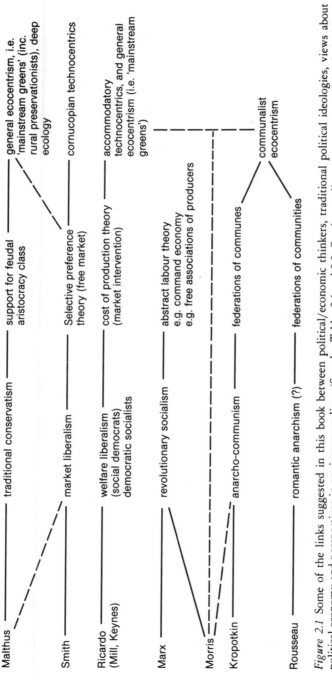

Figure 2.1 Some of the links suggested in this book between political/economic thinkers, traditional political ideologies, views about political economy and perspectives in environmentalism. (See also Tables 2.1 and 2.2. Continuous lines suggest strong links, dashed lines suggest weaker links)

who would freely intervene in nature ('interventionists', who, paradoxically, are non-interventionist in the market economy: e.g. Simon and Kahn (1984)) and those who accept a need to accommodate to natural constraints. Accommodation involves an environmental management approach based on cost–benefit and risk analysis, together with intervention in the economy, via environmental taxes and penalties, standard-setting and the like (Pearce *et al.* 1989). It produces what O'Riordan (1989, 88) calls 'superficially attractive' reforms, and is essentially a survival strategy for the political status quo; a 'whirlpool of contemporary environmentalism into which much intellectual debris is sucked', providing succour for liberal environmental academics and consultants. Less cynically, he makes the important point (p. 87) that interventionists see themselves as environmentalists too:

> Interventionists believe that they can upgrade the quality of existence for all the world's people so long as the right entrepreneurial conditions hold. The quality of life is just as important for them as for the green advocate. The difference lies in the emphasis given to the meaning of that term and the method of achieving the objective. Interventionists see environmental considerations as *incidental* to economic and social advance; green proponents see such considerations as central to their concerns and as the prime objective. Moreover, green advocates fundamentally reject that it is possible to survive through interventionist practices: the Earth cannot absorb the effects of development and people will rebel through 'creative disobedience'.

This last point means that 'green advocates', i.e. ecocentrics, ought theoretically to be concerned with radical social change, away from the kind of interventionism that has characterised liberal industrial capitalism for three hundred years. Marxism and anarchism, being concerned with radical social change, is thus more relevant to ecocentrism than technocentrism. This book mainly addresses ecocentric concerns, expressed through its political ideology of ecologism. It will suggest that these should largely be met through the more anthropocentric radical environmentalism of eco-socialism – except, that is, for the concern not to be anthropocentric, which should be abandoned.

Ecocentrism itself embraces important differences in emphasis within an overall paradigm of nurturing nature rather than intervening destructively in it. There is what O'Riordan (1989, 89–90) calls 'communalism' (see Table 2.1), in which 'economic relationships are intimately connected with social relationships and feelings of belonging, sharing, caring and surviving'. It stems, he says, from nineteenth-century anarchism and seeks to address established socialist principles, in cooperative networks of community organisations. He detects a closing of the gap between communalism and technocentric accommodation.

However, there may at the same time be a widening of the smaller gap between communalism, largely equivalent to 'social ecology', and 'Gaianism' or 'deep ecology'. There has been some vigorous disagreement between these two camps recently, largely provoked by the social ecologist (i.e. anarchist) Murray Bookchin (see Chapter 3.6 and 3.9 and also Bookchin and Foreman (1991) for evidence of some reconciliation). Gaianism combines the old Greek concept of an Earth Goddess with Lovelock's (1989) view of the earth as a complex homeostatic system. This system is immensely resilient, and although humans could destroy it in its present form, it seems more likely that they would destroy themselves and that Gaia would continue – albeit with species other than humans being most dominant. Lovelock's hypothesis does not attribute intelligence to Gaia. But many Gaia*nists* do: particularly deep ecologists and New Agers. Gainism lends itself to New Age mysticism, including paganism, and to the ecocentric *bioethic*, which calls for respect and reverence for nature's intrinsic rights and worth, regardless of use or otherwise to humans. It also demands 'living lightly on the earth', and a deep sense of community involving people and non-human nature.

O'Riordan (1989, 93) says that his classification presents 'a picture of contradictions and tensions dominated by a failure to agree over cause, symptom and action'. This impression will be reinforced when, next, ecologism, the political philosophy of ecocentrism, is considered in the light of the old questions asked by political economy (Chapter 2.2) and is mapped against traditional political ideologies (Chapter 2.3).

2.2 QUESTIONS OF POLITICAL ECONOMY

Schools of thought

Chapter 1.2 described some fundamental questions about the nature of society and how to understand it, which any radical ideology like ecocentrism inevitably becomes involved with. Those questions may be considered broadly 'political' in the sense of their concern with public affairs and how society is run. A further set of fundamental questions arises in connection with, specifically, the running of economic affairs – affairs concerned with how we organise ourselves to maintain our material existence. Clearly, what greens say about such things as consumerism and limits to growth must have profound implications for economics, and for political economy. This latter concerns the moral, political and social desirability of different economic politics (it may be distinguished from 'economics', whose practitioners often – inaccurately – portray themselves as concerned with an objective science describing how and why economies run as they do and forecasting the economic future).

If ecologism, along with anarchism, implicitly says much about political ecology, Marxism is very explicit about it. Hence much of the rest of this book addresses political economy questions as well as those raised in Chapter 1. The eco-socialism presented in Chapter 5 is a brand of green ideology that, like

Marxism, places political economy questions unequivocally at the centre of its concerns.

Cole, Cameron and Edwards (1983) maintain that the development of economies over time reflects the outcome of battles between just three schools of thought on the nature of value and how it comes about. Each school of thought is 'correct' within its own assumptions. There is the subjective preference (SP) theory of value, deriving from political economists such as Malthus, Jevons and Adam Smith. There is the cost of production (COP) theory, owing much to Ricardo, J. S. Mill and Keynes. And there is the abstract labour (AL) theory, which derives from Marx, drawing on some of Ricardo's ideas. Modern Western economies reached a post-World War II 'consensus' around Keynesianism, but from the 1970s there has been a growing split between this COP school and a revived SP 'monetarist' school, spearheaded by Milton Friedman, and put into practice by many governments, notably those of Thatcher and Reagan.

Ecologism in general appears to owe something to both these schools, but little, directly, to the AL school. However, in as much as 'communalist ecocentrism' owes something to the anarchist communism of Kropotkin, as well as to the more liberal and romantic anarchism that is sometimes linked with Rousseau, there are affinities with the AL school too. Each of these theories, and the political policies they imply, is briefly described below (the main sources for this section being Cole *et al.* (1983) and Harvey (1974)). And of course, since Marxism is a central concern of this book, the AL theory is given more attention in Chapter 3.3.

Figure 2.1 is offered as an overview of some of the perspectives discussed in this chapter and elsewhere. It is speculative, in that there might be justified discussion about the links indicated and their strength and importance, and about other perspectives that have been omitted. The figure should be regarded mainly, therefore, as a stimulant to further thought and discussion.

Subjective preference theory

Liberal philosophy underlies this, giving primacy to the individual and legitimacy to possessive individualism. SP starts from the premiss that the value of things springs from individuals who calculate their actions so as to maximise personal welfare. Their individual tastes lead them each to decide between different possible consumption patterns. The range of possibilities is partly determined by the individual's talent, hence the ability to satisfy others' tastes and get an income. The need to do this last springs from the fact that individuals can increase their productivity by joining together in production and applying the principles of division of labour and specialisation, hence there is little incentive for people to try to produce all of their own wants.

So, since different individuals have different demands, according to taste, and produce different things to satisfy others' wants, according to talent, there is a need for *exchange*. This has to be regulated, supply being matched to demand, which is done through a *market*, via entrepreneurs. And, because each person has different tastes and interests, they can be best met through *free* exchange – i.e. nobody has to contract to buy or sell unless they want to for their self-interest. In this way the market is held to reconcile individual interests with those of society as a whole, which is seen as no more than the sum of the individuals comprising it.

Value accrues to a product because someone wants it. But it follows from Malthus's 'population principle' that population constantly tends to outstrip resources. Hence potential or actual *scarcity* underpins all market transactions, and, broadly, the more limited the supply of a good in relation to demand, the higher its price. Malthus's concern with the limits on production led him to allege that an inherent human tendency to breed copiously, displayed particularly by the lower classes, would lead to poverty and famine, with more and more people sharing what was available at little more than subsistence level. He therefore deduced that any wide-scale poor relief administered as of right by government (equivalent to modern state welfare) would be translated by the poor into more people – the better off they were the more they would breed. But increases in resources would be insufficient to meet this extra demand, so the population would soon find itself back at bare subsistence level. Hence the poor relief would have defeated its own object. He therefore argued against such relief.

Modern SP theorists also argue against government intervention in the economy. They consider that economic policy should maximise the freedom in which individuals can exercise their consumption and production choices. And they argue, like Malthus did, that to intervene by providing welfare is ultimately self-defeating. It creates a 'nanny state', making people less self-reliant. Hence government's role should be restricted to removing coercion from the market place, while ensuring that contracts which have been entered into can be and are honoured (this implies creating financial and physical infrastructures to facilitate exchange, and a set of laws, with enforcement, governing exchange). It should also attack inflation, which destroys personal savings and makes the real value of things more difficult to determine in a climate of generally rising prices.

Set against Malthus's concern that the masses should not be given too much wealth, which would result in overbreeding, was his contradictory concern that there would not be enough wealth in the economy for people to buy the things that had been produced. For the working classes were not paid enough in wages to be able to consume most of the fruits of their labour, yet as well as being the producers they also constituted part of the market for their products (the contradiction of overproduction in capitalism is discussed in Chapter 3.3). Consequently the economy would suffer from under-demand, and stagnate. In keeping with his own ideological interests as a defender of the aristocracy, he therefore proposed that it was a good thing that there was an idle rich class.

The aristocracy were conspicuous consumers and so would help to keep the economy going – while not translating their wealth into overbreeding for fear that this would tax it too much, leading to reduced station in life.

Malthus's contradictoriness makes him a problematic figure from the point of view of ecologism. For greens enthusiastically embrace neo-Malthusian limits to growth and overpopulation theses, basing their 'alternative' economics on the same assumption of latent scarcity which underlies free market economics. Yet at the same time Malthus's concern about lack of effective demand led him to preach a capitalistic gospel which has become anathema to greens:

> The greatest of all difficulties in converting uncivilised and thinly peopled countries into civilised and populous ones, is to inspire them with the wants best calculated to excite their exertions in the production of wealth. One of the greatest benefits which foreign commerce confers, and the reason why it has always appeared an almost necessary ingredient in the progress of wealth, is its tendency to inspire new wants, to form new tastes, and to furnish fresh motives for industry. Even civilised and improved countries cannot afford to lose any of these motives.
>
> (from Malthus's *Principles of Political Economy*)

The political ideologies and policies which flow from SP theory (or, to put it more accurately perhaps, which SP theory legitimises) have been developed by many economists since Smith and Malthus (e.g. Hayek). But they are still very recognisable today in Western 'conservative' governments (largely market liberal in reality). Individuals are at the heart of their policies, which include privatisation of nearly all production. Like Malthus, these governments tend to blame the plight of the poor, sick and needy onto the poor, sick and needy themselves. They set up 'markets' in state-run areas of life, like education and health care. Ostensibly they want as many people to hold property and capital as possible. Ostensibly they interfere minimally in the relationship between capital and labour, and they reduce taxes and legal regulations which are held to inhibit free market forces. Environment and health and safety legislation are notable among such regulations.

Costs of production theory

This followed Malthus to the extent of accepting his view that as the supply of subsistence goods rose so would the size of the labouring population grow, but that as more land was brought into cultivation there would be decreasing marginal returns on the investment because rates of agricultural productivity would decline. It therefore starts, again, from the Malthusian premiss that the natural environment sets limits to production possibilities.

Ricardo took these principles and deduced from them a 'pessimistic' view that the free market would inevitably lead to economic stagnation, with most people living only at subsistence level. They would be ranged alongside a small group of landowners, conspicuously consuming economic surplus. But he did not derive from this, as did Marx, a view of society in constant conflict. Instead his desired model was one of equilibrium and harmony – something which could be achieved if economically rational behaviour was maximised. (This idea is echoed in the notion of a pluralist society – see Chapter 1.2.) He was concerned that such equilibrium would partly depend on there being a balance between labour supply and demand (i.e. full employment). But he was also concerned that where this occurred there would be no further opportunity for capital accumulation because of the lack of labour to produce more (leading to stagnation).

The cost of labour is a major aspect of the costs of production in capitalism (raw materials from nature constitute another cost). And it is these *costs of production*, rather than consumers' subjective preferences, which are regarded as the main determinant of what is produced and what is its value. One determinant of the costs of labour is the prevailing technology, which controls the level and sophistication of the division of labour and consequently the amount of labour time required for products, i.e. the labour productivity (as technology advances, the productivity of that labour force which is still left in work generally rises). Labour costs are also a function of how society decides to allocate the wealth from production between labour and capital, i.e. what is considered an adequate material level at which labour should live.

COP theory, as developed for instance, by Mill, is technologically determinist. That is, it largely sees the advance of mechanisation and the division of labour as unavoidable, so that in their working lives most people will inevitably become 'machine appendages'. But in the second respect – in the distribution of the fruits of production – there is room for society to affect and improve the quality of people's lives by determining the levels of wages and profits.

This is an area of relative free will – and of conflicting wills between different vested interests (e.g. unions and employers). The distribution of wealth largely depends on a conflictual bargaining process, and this is what decides whether the given technological innovation will result in higher output or higher unemployment, higher profits or higher wages. This process also influences what and how much is produced, because it influences the costs of production. Because each group in the struggle does have its vested interests, it is necessary for the state to intervene to ensure the maximum of rationality – the state, here, is seen as a neutral bureaucracy.

The political ideologies and policies that follow from all this are mainly those of social democrats (welfare liberals) and democratic socialists. They aim to influence the distribution of wealth so as to let it facilitate the process of dynamic and fast technological change. This means intervention in the economy, to stimulate it in slump, to create jobs and to mediate between different interest groups. Thus government would get employers and workers together to discuss and formulate strategic economic plans. And there would be intervention to provide a 'social wage' through welfare, unemployment and sickness benefits (this is rational, for as well as minimising potential political dissent it provides a pool of potential labour in good health, ready to assist when production can expand). Business and industry would be helped and stimulated to adopt technological innovation in the name of international competitiveness.

COP theorists and practitioners think that there are major structural problems in the free market economy. Keynes noted a long-term tendency to stagnation because he thought that as incomes rise people save more and consume less of their marginal income – here is some idea that there are limits to needs. To this dampening effect may be added the factors which militate against investment in production, such as uncertainty about future revenues and interest rates and corresponding certainty that technological innovation will come thus constantly necessitating the purchase of new machines to maintain productivity.

Because these structural problems are always there, it is argued, then there is always a reason to intervene. The state therefore becomes a social engineer and an economic manager – partly to redistribute wealth towards the deprived who often are so because of the side effects of new technology. The extent of intervention is usually a function of the balance between social democrats and democratic socialists. The British post-war parties have mainly been composed of these kinds of politicians, though in the late 1970s and 1980s the Conservative Party became dominated by the politics of SP: a self-confessed revival of nineteenth-century free-market liberalism, mixed in with some elements of traditional conservatism. But the influence of the AL theory on Labour Party politics has been less notable – after a brief period of domination by democratic socialists with even a few revolutionary socialists in the early 1980s, the party has now slid back to the 'middle' ground of social democracy, i.e. managing capitalism.

Abstract labour theory (see also Chapter 3.2)

This also has a degree of technological determinism, holding that the type of technology determines the technical division of labour through which nature is transformed into products for use. But it also considers that the type of technology, together with the whole structure of production, distribution, exchange and consumption, reflects the relations of production (the relations between groups as a result of their position in the production process). Latent conflict between these groups, which are economic classes, is what determines who controls the means of production and how the economic surplus is to be distributed.

In capitalism, relations of production are expressed in commodity exchange.

Labour is a commodity which is bought and sold. The labour invested in a product ('abstract' labour to be more accurate – see Chapter 3.2) is the source of that product's value as realised in exchange. Some of this value is returned to the workforce so that it can continue to maintain and reproduce itself. The rest is creamed off – appropriated – as capital, and used to invest in further production to gain more capital. The class which does this is able to do so because it owns the means of production, including land and labour (which it has bought).

But there are tensions and contradictions between the technical and social relations of production and these eventually lead to social change. New technology is used in the constant drive to increase productivity and therefore profitability. But (after Ricardo) there are nonetheless long-term tendencies for the rate of profitability to decline. This leads to increased pressures on capitalists to squeeze more out of the labour force, resulting, potentially, in class conflict. This last is regarded as fundamental in capitalism and unresolvable by state intervention. So social progress consists of engaging in the conflict to gain increased control of social relations and the means of production for the exploited classes. Such revolutionary changes, it is hoped, will eventually create a society where there are no longer antagonistic class relations.

The politics which follow from this theory emphasise the analysis of capitalism which shows that it is based on exploitation – i.e. the appropriation by capitalists of some of the value which labour creates. They emphasise that exploitation would not be possible if labour could get control over the means of production, hence this is their main aim. Greater collective control of productive life, through struggle, also depends on much greater democracy and freedom of information. Civil and industrial democracy depends on a degree of decentralisation of power, which facilitates workers' self-management. At the same time the whole economy must be planned and held in common ownership, which for some socialists points to the indispensable role of a centralised state.

Ecocentrism and political economy

There is a clear correspondence between the SP theory and the position of cornucopian technocentrics, such as Simon and Kahn. And COP theory is obviously compatible with accommodatory technocentrism. By contrast, ecocentrism, because it rejects so many of the premises on which conventional capitalist society operates, might be thought not to have much in common with either of these schools. To an extent this is true. Ecocentrism, as will be apparent from Chapter 4, has strong links with anarchism, hence its influences include such figures as Kropotkin and Rousseau. But, reflecting the eclecticism of the ecocentric position, it is also possible to see connections with both SP and COP theory.

Strong neo-Malthusian currents suffuse ecocentrism and its views on economy. There is an almost unanimous acceptance among greens that limits to growth underlie all human activity, and this translates directly into 'Gaian' desires to fashion societies that mirror the rest of nature and are subject to its laws – living within its limits. Atkinson's (1992) pithy summation of bioregionalism perfectly encapsulates this desire: 'The basic idea is to remake our culture to conform to the particular characteristics of land form and ecology in which we find ourselves'.

Within such constraints some green economics favour an SP view. 'If you can't sell it there's not much point in making it' says Richard Adams of the green consumer movement (Hoult 1991, 41). The idea of small communities of 'ethical' businesses is a component of some ecotopias and of some real ecocentric societies (e.g. the businesses surrounding the Findhorn Foundation in Scotland). SP appeals to ecologism's libertarian element (favouring absolute minimal constraints on individual freedom). The primacy which SP theory gives to the individual's role in society meshes well with the green propensity to see responsibility for society and social change starting and continuing with the individual. SP's ostensive dislike of government intervention fits in well with the green critique of contemporary society as overcentralised, bureaucratic and subject to the sinister power of multinationals and state corporations alike.

Cole *et al.* (1983) assert that there are also substantial links between COP theory and the ecology movement. For the technological determinism of this theory reaches into a concern about consumption as well as production. The economists J. S. Mill and J. K. Galbraith saw that what is produced is substantially conditioned by what *can* be produced. And as research and development science and technology come up with new products, these have to be matched by a new demand, created through advertising and marketing. Hence these writers are concerned about mindless and wasteful consumption as an inevitable corollary of mindless, alienated employment.

The political position that might follow from this is ecocentric: industrialisation was a mistake, it has a hidden cost in the loss of essential human values and possible loss of reproductive capacity (through environmental degradation). If the costs of production determine what and how much is produced, then *less* should be produced because the costs are higher than we thought. This implies some planning and conscious manipulation of the market and society.

But such views can also take ecocentrism beyond COP theory, into small-scale communalism. On the one hand this may become an anti-technological, anti-industrial romantic anarchism. Here there are resonances with Rousseau, who was what Cranston (1966) calls a 'normative irrationalist' – that is, he thought that society should not be guided by reason, emphasised the importance of sentiment and natural impulse, and was hostile to civilised sophistication. Rousseau's (1743) *Social Contract* was about political organisation rather than political economy. But he pointed to a form of social and economic organisation in which, he thought, it was possible for individuals to be free and yet be in society. That was the small-scale face-to-face society of Swiss towns and communities, which federated from the bottom up for wider than local purposes. Organised as such, people would willingly give up those lesser 'freedoms' that they might have experienced if they lived in a primitive society for the more prized and civilised freedoms of living under a social contract with other citizens which all freely entered into. The natural rights of the 'noble savage', then, would be exchanged for civil rights, whereby we could be ruled by 'ourselves' – or a sovereign body that truly represented us.

On the other hand there is a green communalism which shies away from the reactionary side of irrationalism, and indeed the possible totalitarian rather than libertarian implications of Rousseau's position (see Russell 1946, 670–4). This is the communalism of utopian socialism and anarcho-communism (Chapter 4), and here is where one finds links with Marxism's AL theory. For the societies which Morris and Kropotkin envisaged were classless, moneyless and characterised by common ownership. The value of things, about which Morris wrote passionately, resided in their *use* in terms of function, ability to please aesthetically or to appeal to the intellect or to the sense of humour: virtually any way, that is, except as items of exchange to realise profits.

Some greens in alternative communities today have an internal political economy which is informed by use rather than value, along with the principle of 'to each according to need: from each according to means'. However, they remain a fringe element that constantly has to compromise with the context in which it is set; that of an international capitalist economy. So the AL perspective does not, as yet, figure more than marginally in green economic theory or practice.

In their review of political economy, Cole *et al.* observe:

> As such ideas [those of Schumacher and other green economists] have roots in a variant of mainstream cost of production thinking, the ideas of the ecology movement seem logically, if problematically, to find expression in social democratic political programmes.
>
> (1983, 180)

The accuracy of this impression is reinforced when mainstream green and green anarchist ideology is compared with traditional political ideologies, as we now go on to do.

2.3 ENVIRONMENTALISM AND TRADITIONAL POLITICAL IDEOLOGIES

Political ideologies

Cole *et al.* take pains to stress that none of the political economy theories which they describe is more 'correct' than any other. Each theory has a band of adherents who believe that it is correct and that the others are 'wrong'. This is because each theory is, in reality, an *ideology*. Marxists define this as a set of ideas, ideals, beliefs and values which derive not from disinterested thought but from the material, vested interests of those who hold them. They are presented as being universally true – as 'common' sense – but really they reflect these more limited interests. Thus the owners of the means of production obviously believe in SP theory – it works to their advantage – while the supporters of labour hold to AL theory. The 'middle class' of managers, professionals, planners and technicians, who are the organisers and facilitators of production, clearly favour COP theory.

Ecocentrics have an affinity with this class, whereas they are suspicious of both capital and labour, hence their frequent penchant for welfare liberal programmes (social democrat), mixed with democratic socialism, which Cole *et al.* refer to. Dobson (1990, 85) has a similar diagnosis about ecocentric political ideology: 'The general aspiration of green ideologies, or the benchmark against which any picture of the sustainable society must be tested is left-liberal ecocentrism'. Table 2.2 portrays most of ecologism as represented by 'mainstream greens' and 'green anarchists and eco-feminists'. It suggests that its political prescriptions (except perhaps the concern for biocentrism) do occupy a spectrum ranging from welfare liberalism to revolutionary socialism. But there are also some aspects of traditional conservatism in ecologism, as Dobson (1990, 31) acknowledges:

> the understanding of the place of the human being in a pre-ordained and immensely complex world in which we meddle at our peril is neverthe- less a right-wing thought. The belief in 'natural' limits to human achievement, the denial of class divisions and the romantic view of 'nature' all have their roots in the conservative and liberal political divisions.

However, before exploring such ideological links further, we must acknowledge that the taxonomy which follows (see Table 2.2 and Figure 2.1) does blur the fundamental distinction which was referred to in the previous section. (As with most classifications, there is also substantial blurring at the edges.) On the one hand are those whose politics claim to be ecology-centred ('ecocentrics', consisting of mainstream greens and eco-anarchists). They prioritise the importance of sustaining 'natural systems' as the *starting point* for their views on social organisation. Theirs are the ostensibly 'new' biocentric politics of ecologism.

On the other hand are those who *start* from social concerns, particularly about wealth distribution, social justice and quality of life. They recognise that the environment is an issue which vitally affects those concerns and vice versa. Hence they wish to assimilate environmentalism into pre-existing perspec- tives. Theirs are the 'old' politics, which are often described in a derogatory

Table 2.2 Political philosophies and environmentalism

Traditional Conservatives (radical)	Market Liberals (reformist)	Welfare Liberals (reformist)	Democratic Socialist (reformist)	Revolutionary Socialist (radical)
Are limits to growth, and enlightened private ownership is the best way to protect nature and environment from over-exploitation. Protect traditional landscapes, buildings, as part of our heritage. Anti-industrialism: human societies should model themselves on natural ecosystems: e.g. should be stable, and change slowly and organically. Need for diversity, but hierarchical structure: bound together by commonly-held beliefs. Everyone to be content with their position (niche) in society. The family (perhaps extended) is the most important social unit. Admire tribal societies. Very romantic: yearn for the past.	The free market, plus science and technology, will solve resource shortages and pollution problems. If resources get scarce, people will supply substitutes – if there's a market for them. Don't believe in 'overpopulation' – people are a resource. Capitalism can accommodate and thrive on protecting the environment. Consumer pressure for environment-friendly products will play a big part, Capital will respond to this market.	Market economy, with private ownership, but managed. Reform laws, planning and taxation for environmental protection. Enlightened self-interest, tailored to the communal good, will solve the problems. Consumer pressure for environment-friendly products will play a large part. Pressure group campaigns, in a pluralist, parliamentary democracy will lead to appropriate legislation.	Decentralised socialism; local democracy; town-hall socialism. Mixed economy and parliamentary democracy – with strict controls on capitalism. Emphasises the role of labour and trade unions. A big role for the state (especially locally). Mixture of private and common ownership of resources. Emphasis on improving the urban environment. Production for social need. Big coops sector. State subsidies environmental protection (e.g. public transport).	Environmental ills are specific to capitalism, so capitalism must be abolished: requiring some revolutionary change, perhaps brought on by environmental crises. Rejects the state ultimately, but perhaps needed in the transition to a communal (commune-ist) society. Class conflict vital in social change to a green and socially just world – reject parliamentary reform. Poverty, social injustice, squalid urban environments, all seen as part of the environmental crises. Similar visions of future to anarchists, but emphasise collective political action, and the state initially.

'Radical' = wanting to go back to the roots of society and change it fundamentally in some ways, and quite rapidly.

'Reformist' = the present economic system is accepted: but it must be revised – in the direction of either less or more interference in and management of the economy – gradually and through parliamentary democracy.

Mainstream Greens (radical aims, but reformist methods)
(inc. British Green Party: Friends of the Earth and other pressure groups)

A mix of welfare liberal and democratic social prescriptions but say they reject politics of left and right. Emphasise the importance of the *individual* and his or her need to revise values, lifestyles and consumer habits. Bioethic, limits to growth, utopianism.

Green Anarchists and Eco-feminists (radical aims and methods)

Advocate a lifestyle of voluntary simplicity. Also, need to change social-economic structures, inc. putting an end to the 'industrial society'. Favour small-scale capitalism, but with profit motive secondary to production for social and environmental need. Also coops and communes. State has a role – especially locally. Romantic view of nature – spirituality important, especially in deep ecology and New Ageism, which all mainstream greens have tendencies towards. New Age irrationalism, mysticism, rejection of 'politics' and industrialism gives it a reactionary, conservative element.

Reject the state, class politics, parliamentary democracy and capitalism. People to organise themselves: have responsibility and power over their own lives. The *individual* very important, but the individual gets fulfilment in relation to the community. Decentralised economy and politics: common ownership of means of production, and distribution according to needs (income-sharing communes). Spontaneous and organically evolving society. Non-hierarchical direct democracy. Rural and urban communes and cooperatives. Bioregionalism.

*These two together represent 'ecologism' (ecocentrism), which starts, unlike others, from the *ecological* imperative and the bioethic (nature as important as human society). But in their social prescriptions they mainly straddle liberalism and socialism (with one or two elements of conservatism).

way by ecocentrics as 'anthropocentric'. They include 'technocentrics' but not all are classifiable as such.

As I suggested above, many ecocentrics might regard this as such an important distinction that they would not be happy to see their politics mapped along the same coordinates as those whose starting point is different. However, because greens do make *social* prescriptions, it can be argued that such an attempt is justified, and revealing. Also, it needs to be remembered that both these groups are 'environmentalists', who would claim to be green in some way. No serious politician, public figure or socially aware citizen in Britain does not claim to be an environmentalist, in the Oxford English Dictionary sense of having concern for the environment, pollution, etc. But ecocentrics consider themselves 'deep' as opposed to 'light' greens, because they are radical as well as bioethical. In fact, when one considers radicalism as opposed to reformism (see Table 2.2), it is clear that this dichotomy is not sufficient to distinguish ecocentrics from the rest, since there are, of course, traditional political philosophies which are equally as radical as ecologism, if not more so.

Traditional conservatism

The words 'conservation' and 'conservatism' both have the same root. And in traditional conservatism the ideas of tradition, continuity, stability and dislike of sudden change but an acceptance of slow organic change (i.e. which is not planned or blueprinted or revolutionary) – all of these are compatible with some environmentalist thought. In conservatism the analogy between society and nature is strong: just as ecosystems need to be changing *organically*, not precipitately, so does society. As with nature, variety and structure in society are essential to achieve the all-important goal of stability. Social revolutions (and new technology) upset the natural social order. This order is perceived to be hierarchical, though each link in the chain of being is worthy of some respect. Hence relationships between higher and lower orders might be oppressive and economically exploitative but they nevertheless involve mutual obligations: they are not just one-way. Those lower in the social order should accept the naturalness and inevitability of their position.

These beliefs foster a romantic view of 'traditional' societies. So also is there a liking for 'traditional' pastoral landscapes and grand architecture, expressed in Britain in the work of long-established nature and conservation groups like the National Trust, Councils for the Protection of Rural England and Wales, County Landowners' Association, Farming and Wildlife Advisory Group and the Civic Trust. Conservative environmentalism frequently takes the form of a conception of 'stewardship' on the part of landowners – holding the land in trust for a future generation (again there is the notion of obligations attached to power). Hence, the Bow Group, a right-wing pressure group of the Conservative Party, emphasises Edmund Burke's injunctions about wise stewardship, and proposes traditional virtues of efficiency, order, thrift, self-help, tradition, patriotism and nationalism as the basis of Tory green-ness (Paterson 1989).

Conservatism agrees with the message behind Hardin's (1968) 'commons' parable, arguing that enlightened private ownership of resources is the best way to conserve them. It might also accept Hardin's (1974) arguments for coercion to curb population growth, and to promote the 'correct' social and environmental consciousness.

As Hay (1988) points out, traditional conservatism may grade into fascism, also known as 'right-wing irrationalism', or 'extreme' romanticism. Its key elements include the use of biological metaphors, the stress on the organic community and the individual's need to merge with it, the elevation of ritual, intuition, and the mystical, and the distrust of the rational. These elements may all be found in some ecological writings, especially those promoting deep ecology activism. Bookchin (1986a) warns:

> I have been a student of nature philosophy all my life. I have seen how nature philosophy can be gravely abused. For example, it is only too well known that biological explanations or even ecological explanations have been used to support fascism. Hitler, for example, used biological analogies, notably those of race, soil, homeland, folk, blood, to underpin his viciously fascist imperialistic theories. He spoke about the homeland almost in an ecological sense.

However, Bramwell (1989) is wary about applying the term 'ecofascist' to anyone in the green movement. She believes it is better to reserve the term for the real fascist movement. She notes that neo-fascist organisations in Europe (Germany, France, and since 1984 the National Front in Britain) have all taken onto their platform a 'green' perspective. And she also describes how German, as well as other European Nazis, showed a liking for vegetarianism, the 'back to the land' movement and bio-dynamic farming.

Market liberalism

Sometimes called nineteenth-century free market liberalism, or neo-conservatism, this ideology is championed by technocentric 'cornucopians', following the example of Simon and Kahn (1984). They are aggressively optimistic about the potential of the free market allied to technology to solve our environmental problems. The invisible hand of market forces under which individuals pursue self-interest, they argue, gives society more environmental protection than will any kind of intervention or regulation, which is a constraint on liberty. Thus if a 'natural resource' is running out, its increased scarcity will push up the price of the goods or services that come through that resource. This will encourage entrepreneurs and scientists/technologists to try

49

to devise some substitute or more ingenious ways of providing the same goods and services. Similarly, there is money to be made from non-harmful aerosols, biodegradable plastics, catalytic converters and the like, so there is no practical dichotomy between the interest of capitalism and environmental quality (Elkington and Burke 1987).

Welfare (ameliorative) liberalism

Such arguments are poorly supported by the actual environmental record of *laissez-faire* governments and industry over the past decade. Hence welfare liberalism is coming rapidly into fashion. Welfare liberals also believe in capitalism, but not without restraints and controls to limit its harmful effects on people (the economic losers) and the environment. They emphasise the role and supreme importance of the *individual*, and his or her enlightened self-interest in protecting the environment. In a pluralist democracy, such as (they believe) we live in, parliament is the main forum through which environmental views and interests will be heard and protected. And rationality, the rule of law, technology and environmental and economic management (cost–benefit analysis, reform of taxation) will all help to secure the goals of environmentalism. Following the 'father of English liberalism', John Stuart Mill, they are ambivalent about how desirable unselective economic growth is, and appreciate the need for diversity in society and nature (compatible with their belief in pluralism). Support for private ownership of resources, and the notion of the invisible hand is still strong. But they are also aware of the wider society and communal good, and that what is good is what brings most benefit and happiness to most people (Bentham's utilitarianism). By this measure environmental protection is rational and desirable. Hence planning laws, taxes on non-recycling industries or pollution, and welfare provision to enhance urban environmental quality and environmental education are all legitimate ways for the state to intervene, but only as much as is strictly necessary, in the free market.

Welfare liberals are technocentric 'accommodators' and generally technological optimists, for example Pearce *et al.* (1989), who wrote a report that is likely to influence British environmental policy for some time in the future, arguing, as it did, for tax incentives and disincentives, permits and other semi-coercive management devices to control pollution and resource use. Much of the eco-consumerist movement (Ark or the Body Shop, to which Greenpeace and Friends of the Earth are guardedly sympathetic), is welfare liberal, as, of course, are local Liberal Party activists in Britain, who have, since the last war, been among the most environmentally conscious party political members (especially when you include as environmental issues matters like housing quality, provision of schools and social services).

Democratic (ameliorative) socialism

Including, as it does, part of the decentralist tradition in socialism, this can shade into anarchism. It also, however, believes in the need for pluralist democracy and the power of parliament as one way to achieve social change, and that before capitalism can be abolished there may need to be a phase of 'managing capitalism' by the *state* to achieve desirable social and ecological goals – so in this way it shades into welfare liberalism. Class analysis, especially in Marxist terms, is muted, but collective political action is seen as important, though the near-primacy of the individual is acknowledged.

Ultimately, but not very loudly, this form of socialism is opposed to the capitalist principle of production primarily for profit, in favour of production for social and environmental *need*, perhaps alongside profit. It does not reject industrialism but criticises its capitalist form and encourages common owner-ship of the means of production (though not necessarily distribution purely according to need). Cooperatives constitute a major way of organising production in this socialist ideology, along with forms of local community organisation and local democracy, administered and facilitated by municipal socialism. The 'loony left' local councils in Britain – Sheffield, Liverpool, Greater London Council, for example – are and were not loony from a green socialist point of view, and have done much for environmental causes, from recycling to improving the welfare of minority groups and women. In this socialism is implicit the view that 'the environment' must be more widely defined than just connotations of 'nature' – a point of view reflecting the urban base of socialist history. So there is a big role for the *state*, as a facilitator for local autonomy and a partly decentralised society. Ryle (1988), Frankel (1987) and Stretton (1976) reflect this perspective on environmentalism, as does SERA (Socialist Environment and Resources Association), an environmental lobby group attached to the British Labour Party, and the Association of Socialist Greens, a lobby in the Green Party.

Revolutionary socialism (see Chapter 3)

This shares with democratic socialism the fundamental rejection of capitalism (but is more strident about it), the emphasis on production for social need and environmental quality, the acceptance of the state (at least in the transition to commune-ism), and the need to define 'environment' in social as well as natural terms. The analysis of the society–nature relationship generally follows Marxist lines, which see environmental problems as inherent in the nature of capitalism. However, opinions differ on how much they may be *solely* located here (see Grundmann 1991). The analysis of how to get to an 'ecological' society, which equates with a – perhaps moneyless – communist one, is based on the mechanism of class struggle. Neo- and orthodox Marxists (Chapter 3.1), however, may differ on who might be the principal agents and

51

actors – faith in the potential effectiveness, or even existence, of a working-class proletariat is not universal (Gorz 1982).

Marxists tend to see the rise of environmentalism itself in class terms – many ecocentrics being thought of as bourgeois and counter-revolutionary. This may mean open hostility towards environmental groups and campaigns, though more often than not there is an uneasy alliance between reds and greens based on agreement on *ends* but not necessarily on *means* or analysis of *causes*.

Marxists by and large believe that environmental, feminist, peace and third world campaigns are, or should be, all part of the ultimate struggle against global capitalism itself. From their materialist perspective the ills which these campaigns highlight are all outgrowths of capitalist relations of production. Such campaigns, furthermore, should focus less on the reform of the individual's attitudes and values; more on the collective political struggle of the world 'proletariat'.

André Gorz and Rudolph Bahro have tended towards the revolutionary socialist view in the past, though they no longer do so. The SPGB, Socialist Workers' Party and Militant are all British political groups who incorporate green issues into orthodox Marxist analysis, while some academics, like Grundmann (1991) or Atkinson (1991) take a neo- if not to say post-Marxist perspective on them.

Mainstream greens

This group of ecocentrics, in the mainstream of ecologism in Britain, is inspired by writers and activists like Porritt (1984), Schumacher (1973), Schwarz and Schwarz (1987) and so on. They particularly include environmental campaigners from such groups as Friends of the Earth and Greenpeace. Many favour the 'reform-of-lifestyle-and-values' approach, combined with pressure group politics founded on the liberal assumption of a pluralist democracy. They also constitute the majority of the British Green Party and dominate green economic thinking (e.g. Ekins 1986, Dauncey 1988, Robertson 1990). As suggested above, their political approach is pragmatic and eclectic, and most of them would probably be pleased to know that it is not easy to pigeon-hole them into old political categories.

They hold it as a point of principle that they are neither left nor right, but 'forward' or 'above the old politics': 'The basic political choice today is not between Right, Left or Centre, but between conventional grey politicians and the Green Party', said that party's 1992 manifesto. But of course, this rejection of traditional politics and politicians can in fact be thought of as a fundamentally conservative ideology. And when they start to talk about what we should all *do* about eco-crisis, greens do invoke the 'old' politics. These seem mainly to straddle the categories of (a) welfare liberalism and (b) democratic socialism.

Thus greens say that social change must proceed from individuals (a) but change is also needed in the economic structures of society (b). They do not totally reject capitalism – indeed are enthusiastic for at least small-scale versions of it (a), but they see social need and environmental quality as criteria to be elevated above the profit motive (b). The state does have a benign role (b), in facilitating the development of individual responsibility (a). This grudging acceptance of the state (and of parliamentary politics by Green Party supporters) constitutes a major distinguishing feature from eco-anarchism. The elevation of natural laws and ecological principles marks mainstream greens off from 'straight' liberals and socialists, as does their sometimes expressed desire for more urgent and radical social change. Nature may be the source of social laws (a) but, to many, principles of social justice are important (b): however *eco*-centrism purports to make social justice part of a wider justice required for all life forms, whereas when the chips are down eco-socialists put human interests first (Eckersley 1992, 128). Technology is not rejected, but it must be appropriate and democratic, as well as 'soft' on nature. Rationalism (a and b) must be balanced by elevating emotional and intuitional knowledge. Democracy and individual freedom (a) are cornerstones of mainstream green ideology – and that democracy is to be extended to all nature's creatures (animal rights, vegetarianism, veganism). But the import-ance of the community is stressed too (b), though not very often of the collective in the production processes (e.g. trades unions).

But besides all this there is also that rejection of the 'industrial way of life', and 'the old politics', coupled with a tendency towards irrationalism and mysticism which is particularly evident in deep ecology and New Age approaches. Whether they publicly own or disown this spiritual wing of the radical ecology movement, it is apparent that most mainstream greens do have deep ecology/New Age 'tendencies' (Pepper and Hallam 1989). This shows in their support for a bioethic, for nature mystification (Gaianism), and their belief in spiritual paths and self-discovery, self-realisation and consciousness raising through therapy techniques as routes to political change.

The innate conservatism in such idealisation of nature and in the denial of a politics of social change in favour of one of individual change is clear. Indeed, it cannot be denied that despite the *emphasis* on left-liberalism in ecologism there is also a persistent strand of conservatism. It may be a minority strand, which is why Table 2.2 does not extend 'mainstream greens' over towards the conservative side of the diagram, but it is there. The most prominent British radical ecocentric associated with it is Edward Goldsmith (1988), who argues for commonly-held belief systems – such as those enshrined in strong religions – as stabilising forces to create that *social unity* which he considers to be a key feature of an ecologically sound society (everyone must believe in the primacy of ecological laws). Goldsmith (1978) has held up the oppressive caste system in India as the kind of social organisation which is compatible with an ecologically and socially sound society – sound because it is stable and therefore deemed to be 'in balance' with the natural environment. The way of life, like the structure and mechanisms of an ecosystem, is designed to maintain order. For Goldsmith the common values must be, above all, those which stem from the ecosystem's model of society – they are not arguable, they are *absolute*. He also argues for small-scale organisation as the geographical basis for achieving ecologically desirable things. This desire for 'traditional' values leads him constantly to refer to 'primitive' peoples and tribes in Africa, Australasia etc., as models for us. And he argues from the assumption of the family as the essential unit of social organisation: whatever preserves this (such as the traditional stereotyped role for women) is to be encouraged. Finally, as a conservative ecocentric, he rejects industrial society – it is aberrant.

Green anarchists (see Chapter 4)

These ecocentrics include 'eco-anarchists', 'eco-feminists' and 'eco-pacifists'. They all share fundamental beliefs about the need for 'organic societies' but some are anti-urban and anti-industrial, so display affinities with conservative thought. Many have decidedly liberal leanings, with an idealist approach to social change, and a rejection of class analysis and of any possible role for the state. Above all, in espousing the individual as the basic social unit, anarchism is sometimes thought to embrace the cornerstone of liberalism: although the anarcho-communist and anarcho-syndicalist would define the concepts of the 'individual' and individual freedom in a way different from that of a liberal.

But then anarcho-communism and anarcho-syndicalism have many affinities with socialism, and it is probably true that many *eco*-anarchists are mainly anarcho-*communists*, looking particularly to Kropotkin for inspiration. There is the rejection of capitalism, desire for common ownership of the means of production (resources), and, in resource- and income-sharing communes, distribution according to need.

The eco-anarchist's utopia may involve rural communes and the craft-based socialism of William Morris's *News from Nowhere*, for example the 'Green Anarchist' group based in Oxford. But it may also be inspired by urban anarchists like Colin Ward, and be based in urban communes and the squatter movement ('property is theft'). However, Anglo-European eco-anarchism does not seem generally to follow the lead of Australian anarcho-syndicalism, which is uncompromisingly urban-centred, rooted in trade unionism, and has been a powerful force in green activism.

This is perhaps because British eco-anarchists have some major divergences from socialism. For instance they generally reject class politics, seeing social change as consequent on the action of individuals in forming spontaneous, mutualist, non-hierarchical groups to *live out* their politics (e.g. in communes), and set an example for others to follow. The 'personal is political', and 'person equals planet' are maxims from more liberal American anarchism (i.e. Roszak 1979) which appear to have great influence.

Indeed, the British eco-anarchist's vision of an ideal society, involving small-scale, collective, decentralised commune-ism, participatory democracy, low-growth (or no-growth) economy, non-hierarchical living and consensus decisions and the rest – all this is highly coincident with the American picture presented in Callenbach's (1978) novel *Ecotopia*. Eckersley (1992, 145–70) subdivides eco-anarchism (see Table 4.2) into the 'social ecology' of Murray Bookchin, 'bioregionalism', 'eco-communalism' and a more spiritual form of the same, termed 'monasticism'. In this book I discuss the first two of these more than the last two, which I have dealt with more fully elsewhere (Pepper 1991).

2.4 GREEN POLITICS ARE POSTMODERN POLITICS

Postmodernism

Many of the issues discussed in Chapter 1.2 were presented as conflicts or choices between opposites, i.e. as dualisms. And it was essentially a discussion of the relative merits of grand theories, attempting to explain societies and social change in terms of universal principles. The same applies to the political economy theories described in Chapter 2.2. This approach – thinking in dualisms, and searching for overarching theory – is all in keeping with a modernist perspective. 'Modernism' is a word describing the tenor of much Western thought over three hundred years, originating in philosophers like Descartes, Locke and Kant, and believing in reason, science and progress:

> what Habermas (1983, 9) calls the *project* of modernity came into focus during the eighteenth century. That project amounted to an extraordinary intellectual effort on the part of Enlightenment thinkers 'to develop objective science, universal morality and law, and autonomous art according to their inner logic' . . . doctrines of equality, liberty, faith in human intelligence (once allowed the benefits of education) and universal reason abounded.
>
> (Harvey 1990, 12–13)

Modernism has promoted the development of 'rational' forms of social organisation – in practice the spread of capitalism – searching to use technology which exploited the laws and resources of nature to produce goods for mass markets; ostensibly to improve universal human wealth and welfare. Until recently the highest development of this was 'Fordist' production (large scale, centralised, production line, division of labour, standardised products, 'scientific' management (Taylorism) and mechanisation). And everywhere there has been in practice constant large-scale development and planning, then renewal, upheaval, innovation and discontinuities. This has been complemented by a search for underlying order and for the meaning of it all by social, economic and historical theorists and by artists, planners and architects.

A desire to recognise the underlying order and structures of this constantly changing pattern has underlain their search.

Modernism, then, has involved a continuous process of destruction of what went before, in pursuit of general principles that were thought to have been desirable for universal human good, for instance freedom from material want, and freedom to accumulate wealth. Many have seen this process as creative. But there has always been a countercultural current which emphasised its destructiveness, questioned its notions of progress, and bemoaned how it demeaned and downplayed other cultures, value systems and points of view.

Over the past twenty or so years such negative views about modernism have grown, contemporaneously with new social movements (such as the greens), and with a new, Rousseauvian, regard for non-rational thought and for other cultures and points of view and for eclectic styles and outlooks. Accompanying this, some think, has been 'post-industrialism', a form of 'flexible' capitalist accumulation involving globally decentralised and smaller-scale production along lines which are thought to repudiate many Fordist principles, but offer less secure and more fragmentary employment patterns and experiences. The organised capitalism which produced the upheavals and modernisations of yesteryear and was theorised by neoclassical economists, or by Marx, has supposedly been replaced, through a dramatic, epochal change, by a new disorder of self-perpetuating consumption. Correspondingly, the academic search for overarching theories ('meta-theory') to explain the world, or universal ethics by which to organise it, has become less fashionable. The belief is about that what we see – surface appearances – *are* the only reality. Through Nietzsche, Husserl, Heidegger and others, the subject–object distinction, the penchant for dualistic thinking, the belief in an objective world and the virtues of rationality – these cornerstones of modernism – have become suspect.

All of these trends in Western society, economics, art, architecture, philosophy, sociology and so on, have been reviewed by Harvey as facets of a 'condition of postmodernity' which, he says, is symptomatic of late capitalist development. (That you *can* pull together such diverse elements and explain them all by overarching ideas of a single postmodern condition is in fact a modernist conception, and the attempt does not go unchallenged – for example by feminists (Morris 1992).)

Postmodernism, Harvey says, rediscovers the vernacular in architecture, emphasises discontinuity in history, and indeterminacy in science (e.g. chaos theory), and the validity and dignity of all possible views and perspectives in ethics, politics and culture. It respects, therefore, the 'otherness' of different perspectives and cultures within the Western world and from further afield. It is therefore 'culturally relativist', but in a sense that modernist ideas of cultural relativity (e.g. as found in Marxism) are not. For while modernist philosophies may understand that different cultures have different world views, conventional wisdoms, common senses, and ethics and morality – that all make sense within themselves – such philosophies do not see them as all *equally* valid in comparison to their own 'modern' values. Postmodern cultural relativism, however, does equally value them – a syndrome sometimes caricatured as 'anything goes'.

And postmodernism holds that the surface appearances of a world increasingly experienced via sound reproduction and pictures and images, where the main use value of many consumables is in creating or enhancing our individual or our group status, *is* reality for most. So postmodernism celebrates surface and superficiality, style, ephemera and consumerism. It has been attacked as an 'intellectual' gloss for Thatcherism, and it certainly seems to be more consonant with late capitalism than a reaction against it. Postmodernism's world is fragmented, ostensibly lacking order and sense of direction. Its irrationality undermines belief in linear reasoning and progress – and it holds that there are no underlying structures from which to read off consciousness, culture and politics (Scott 1990, 103).

Green postmodernism

Green politics often lack structure and coherence, reject authority and embrace cultural relativism – paradoxically despite their desire to see *all* societies conforming to universal meta-theories of ecology, i.e. the laws of nature like carrying capacity. Therefore green politics have much in common with postmodernism. They reject universals (apart from laws of ecology) being imposed on groups, in favour of self-determination, and they reject, in green theorising, the hidden and structural in favour of the superficial. Hedonism (doctrine that pleasure is the chief good and proper aim) and aestheticism (appreciation of beauty) rather than grand morality, are the organising principles behind many visions of a green ethics and society. Atkinson (1991, 61–2) encapsulates green postmodernism. He advocates relativism, believing that to champion universal rationality and dualistic, reductionist, analytic thinking amounts to a millenarian, self-righteous cultural *imperialism*. For since it does champion these things

> Social science, whether positivist or Marxist, as ideological adjunct to this social and political system [a power hierarchy] acts as legitimation of instrumentalism. Its 'discovery' of the 'function' behind non-instrumental cultural manifestations represents a simple *hegemonic denial of the validity of other cultures* or non-instrumental cultural attributes [emphasis added].

Postmodernism, by contrast, is keen to acknowledge the equal value of other cultures and ideas – of 'otherness'. And it is clear that Atkinson's 'radical' relativism is in fact very similar to *anarchism* in its fundamentals. Its mission is to criticise and reject the Enlightenment project, and to create an alternative political ecology: one which will reconstruct society out of the 'manifest disintegration of cultural values characteristic of the "postmodern condition" (ibid., p. 75).

The red–green debate

Seen from the perspective of Table 2.2, the red–green debate in Britain takes place largely between those in the liberal middle of the spectrum of ecologism and those more towards the revolutionary socialist end. It is, too, joined by revolutionary socialists from outside ecologism. From the perspective of the last chapter the debate is a struggle between the abstract labour theory of value (socialists) and the eclectic ecocentric mix of cost of production theory, Rousseauvian romantic anarchism and other influences. But it can equally be regarded as a debate between people who are, for the purpose of settling social issues, in the two camps of modernism (red) and postmodernism (green).

Ecologism (mainstream as well as an overtly anarchist version) is highly infused by elements of anarchism, which has many coincidences with postmodernism, even though it is an old political philosophy. The red critique of ecologism is an attempt to pull it towards a more modernist outlook, involving: (i) a form of anthropocentrism; (ii) a Marxist-informed (materialist and structuralist) analysis of what causes ecological crisis; (iii) a conflictual and collective approach to social change; (iv) socialist prescriptions for, and visions of, a green society. These last include, from the revolutionary socialist perspective, anarcho-communism and -syndicalism, and, from the more reformist socialist viewpoint, local community and municipal socialism.

At the same time, the old struggles and differences between anarchists and socialists, which dogged left internationalism in the nineteenth and early twentieth centuries, are still germane, in the red–green debate. Similarly, Marx and Engels's critique of petty-bourgeois utopianism is relevant in any critique of ecologism from the left, for, as Hulsberg (1985, 17) points out, the formal programme of the greens is a variation on such utopianism.

We go on now to explore the position towards which reds want to pull the greens. It is heavily infused by the abstract labour theory of value, therefore Chapter 3 is broadly structured in two parts. After a brief discussion of the relevance of studying Marxism, Chapters 3.2 to 3.5 outline Marxist political economy and its direct implications for environmental issues in a capitalist society. Then Chapters 3.6 to 3.9, still informed by Marxist political economy, describe a green socialist perspective – its view of nature, its vision of freedom, strategies of how to get there and, finally, its critique of the ecocentric approach. This opens the way to consider, in Chapter 4, anarchism and its influences on ecocentrism. Chapter 3.1 makes it clear that the interpretations of Marxism and socialism in this book are not the only possible ones. But they are the ones that are most consistent with eco-socialism: the next stage towards which green ideology should develop.

3

THE MARXIST PERSPECTIVE ON NATURE AND ENVIRONMENTALISM

3.1 MARXISM'S RELEVANCE TO ECOCENTRISM

What is Marxism?

Marxism is several things. Many people regard it as a political doctrine, although regimes which are dubbed 'Marxist' – self-styled or otherwise – are often gross perversions of the political philosophy of socialism to which Marxism is committed. This is not unusual: National 'Socialism' was in fact the antithesis of socialism, just as many organisations and countries now called 'Christian', 'Free', 'Democratic' or 'Communist' are frequently anything but these things. You cannot judge Marxism, socialism, or any other world view solely by the actions of those who profess them.

Marxism is also sometimes called a 'philosophy', although by denying the usefulness of purely abstract thought and advocating a dialectic between thought and action it could be thought of as 'anti-philosophy' (Marx said 'The philosophers have only interpreted the world in various ways: the point, however, is to change it').

Above all, Marxism is a Western intellectual tradition inspired by Marx but developed by many others. It tries to analyse how society 'works' and how it changes. It is particularly interested in the change from feudalism to capitalism, how capitalism functions, and how it will probably cease to function, perhaps giving way to socialism and, ultimately 'true communism'. Marxism is often written in difficult, unapproachable language. Different parts of Marx's own writings say different, sometimes contradictory, things. Marxism's adherents have, in practice, sometimes committed repression in its name. And because it has a clear moral commitment to socialism and a devastating critique of capitalism it threatens many entrenched interests in the West. So it is unsurprising that Marxism's critics are many – ranging from the right-wing tabloid press to left-wing intellectual 'post-Marxists' – and including many mainstream greens and green anarchists. But to infer from this that Marxism has had its day (as, for example, Porritt and Winner (1988) do) is hugely misguided. As Bertrand Russell (1946) pointed out, various features of Marxist analysis have proved so useful and telling that to an extent all of us in the West are 'Marxists' – we have taken on board elements of a Marxist perspective in our way of thinking (especially its materialist perspective).

Marxism: ecocentric or not?

'Marxist theory has not yet been grounded in ecological science, despite efforts from socialists from Podolinski to Commoner to persuade Marxists otherwise', says O'Connor (1991c). He notes that traditional accounts of historical materialism emphasise how humans transform nature and downplay nature's effects on humans and the rhythms of nature's economy.

Deleage (1989) thinks that there is in fact a total concept of the society–nature relationship in Marx, but that in concentrating his analysis on the capital–labour relationship, Marx lost an opportunity to explore it. And it remains largely unexplored by Marxists. He accuses Marx of asserting (in *Grundrisse*) that capitalism could emancipate itself from natural limits, of ascribing no intrinsic value to natural resources (labour being the only source of value) and of totally neglecting energy balances in his descriptions and evaluations of the production process. Martinez-Allier (1990) adds that although Engels was interested in energy flow and the second law of thermodynamics, he rejected energy accounting in economics, as too difficult. Hence, because of all this, and the 'metaphysical status' which they accord to the production process, Marxian economics are no different from mainstream economics, and use the same language as capitalism. Neither Marx nor later Marxists considered how an economy (capitalism) based on using (allegedly) exhaustible resources might use up the means of production. Furthermore, 'A preoccupation with the intertemporal allocation of exhaustible resources is generally absent from Marxist economics, and this is not because the problem did not exist before 1973' (ibid., 220). Martinez-Allier believes that there is no school of Marxist ecology because Marx's view of history (i.e. in the *Critique of the Gotha Programme*, 1875) envisages *unlimited* development of productive forces under socialism. A Marxist discussion of communism without growth is thus still pending – but, as Deleage puts it: 'One must ask questions about the physical limits to growth and more profoundly about the entropic nature of all economic activity'.

These criticisms sound a little as if they come from mainstream greens: they seem to be built on premises that there *are* limits to growth and resource availability, and that energy budgeting should be paramount in economic analysis. From them to Fry's (1975) ill-judged attack on what he supposes to be 'Marxism' – in fact Stalinist agricultural collectivisation in Russia – is not such a long step. Hence it is interesting to realise that all the critics cited above, except the ecocentric Fry, are Marxists.

They are among those Marxists who place, as Redclift (1987, 48) put it, 'The Promethean quality of early Marxism in doubt'. (Prometheus was fabled to have taught humans how to transform the earth by fire and other 'arts').

Redclift gives the example of Ensensberger (1974), who, in a famous early Marxist critique of ecologism, declared that future world problems would be about survival rather than (as orthodox Marxists might say) how to distribute abundance – an insight which 'showed some prescience. . . . After the recurring crisis of African famine and the nuclear catastrophe at Chernobyl . . . [this insight] looks increasingly realistic'. Redclift thinks that Marx and Engels overemphasised the role of production (especially commodity production) in determining how we use and conceive of nature. While this is important, other areas of life – such as processes of 'biological and social [rather than economic] reproduction' – are equally important in this respect. He takes Marxism's neglect of feminism as an indication of this wider neglect. He also supports Sayer's (1983) objection to Marxism's emphasis on how 'social forms', including economics, mediate nature. They *may* do, but this does not mean, Sayer thinks, that nature is therefore *reducible* to social forms, which may be an implication of the dialectical view of nature (see Chapter 3.6). Hence, there now seems to be a school of Marxists who accept the green notion of significant limits placed by nature on human activity (especially economic), and think that Marxism should be modified to reflect this 'ecological' perspective – a perspective which often says that both socialism and capitalism are infused with notions of growth, and are, in this respect, as bad as each other (see Milbrath (1989) for such a typically ecocentric critique).

Others, however, appear to believe that Marxism does already contain enough in the way of a meaningful – albeit mostly *implicit* – perspective on ecology. Parsons (1977), for instance, is sanguine about 'Marx and Engels on ecology'. He concedes that in their sometimes 'unguarded' statements they said little about nature's value independent of human needs and purposes, and that they shared the general nineteenth-century optimism in material progress and evolution (see Oldroyd 1980) which led technocentric Victorians to reject Malthusian 'limits'. But he believes that the notion in Marxism of advanced society's 'mastery' over nature does not in fact imply a despotic master–servant relationship so much as a skill and intellect that gives an *ability* to transform nature wisely in pursuing legitimate human need. Thus Parsons insists that

> on the question of mastery, Marxists must continue to make clear, as Marx and Engels did, that their ecological position is the very antithesis of capitalism: governed by care and not avarice . . . generous and not possessive, planful for nature and society.
>
> Parsons (1977, 70)

At worst, their attitudes to nature constituted an open question: at best they had an understanding of an ecological approach to the society–nature relationship which predated that of Haeckel (who coined the term 'ecology') himself. For to them

> man is inconceivable apart from his evolution in nature and his collective

labours upon nature by means of his tools. Man's dialectical relationship with nature, in which man transforms it and is therefore transformed, is the very essence of his own nature . . . nature is definable as the materials and forces of the environment that create man and are in turn created by man.

(Parsons 1977, xi: see section 6 below)

Furthermore Marx and Engels 'had a definite (though not fully detailed) ecological position. As both working people and nature are exploited by class rule, so they will be freed by liberation from class rule' (ibid., xii).

Parsons believes that their ecological position comes through via their writings on society's interdependence with nature and the mutual transformation of humans and nature through labour, and through their views on technology, precapitalist society–nature relationships, the capitalist ruination (alienation) of nature and people, and the transformation of that relationship under communism.

Parsons is by no means alone in such views. Vaillancourt (1992) analyses a range of works: *Early Writings, Economic and Political Manuscripts, Capital, Anti-Duhring* and Engels's *Dialectics of Nature*. From them he concludes that Marx and Engels were *forerunners* of human, political and social ecology. They were especially sensitive to the interdependence of humans and nature: their materialism sensitising them to the importance of the natural environment as part of the productive forces, and their humanism highlighting socio-economic influences on nature. Hence they went back and forth between anthropocentric (human centred) and naturalist perspectives (the latter see mind as dependent on material nature, and not in some way prior to or more real than it), having been much influenced by contemporary writers on biological and human ecology. They may have acknowledged capitalism's immense power to develop productive forces, potentially freeing humankind from the 'realm of necessity', (i.e. from living at the mercy of unmodified natural cycles and limits). But they thought that 'Despite certain progressive aspects, capitalism . . . both dehumanises man and perverts the natural world' (Vaillancourt 1992, 34). They favoured active and planned intervention in nature but not its triumphant and ultimately irrational destruction.

We will explore such views and their interpretation by Marxists and Marx-influenced socialists – not least among whom was William Morris, who, 'in many of his essays and lectures, sketched out the principles of what today would be called an ecological society' (O'Sullivan 1990, 169). Morris thought that a society liberated through socialism would think about art, and about making attractive urban environments. He also explored 'green' themes like simple lifestyles, harmony with nature, the inherent wastefulness of the market, and, above all, the need for 'useful work versus useless toil' (1885) that would produce useful and beautiful products, mental and physical pleasure, and a revival of creativity. All this would be in small workshops, and via selectively applied division of labour (Chapter 4.4), where science and technology have saved people from arduous and unpleasant work, and have taught 'Manchester how to consume her own smoke, or Leeds how to get rid of its superfluous black dye without turning it into the river . . .' (*Collected Writings*, cited by O'Sullivan, p. 171).

A sense of community underlies all Morris's vision: he was an 'ecocentric' of the communalist rather than Gaian type (O'Riordan 1989), and what he did, says O'Sullivan

> was to take Marxism and apply it to the practical realities of everyday life. . . . What he also achieved, by no means incidentally, was to provide radical environmentalists with a document setting out many of their basic ideas in plain English [making] an unrivalled contribution both to revolutionary thought and to environmentalism.

Why Marxism is useful to ecocentrics

First, Marxism reminds us that for most people, nineteenth-century environmental problems were *clearly* socially inflicted, through economic exploitation associated increasingly with urbanisation and capitalist industrialisation (including industrialisation of agriculture). This is still substantially true today, world-wide. Thus for Marx and Engels, 'The primary places at which ecological damage was inflicted were the factory and dwellings of industrial workers, the large agricultural estates and rural slums' (Parsons 1977, 22). And in describing the *Conditions of the Working Class in England* in 1844, 'Engels set forth the ecological dislocation produced among the industrial, mining and agricultural proletariat by their urban and natural environments', while Marx, in *Capital* and *Grundrisse*, described how in factories

> Every organ of sense is injured in equal degree by artificial elevation of the temperature, by the dust-laden atmosphere, by the deafening noise, not to mention danger to life and limb . . .
>
> (*Capital*, Vol. 1, 422–7)

Indeed, millions of people experienced an environmental crisis in nineteenth- and early twentieth-century Britain, and books like Tressell's *The Ragged Trousered Philanthropist* and Greenwood's *Love on the Dole* were effectively environmental protest books, while the trades union movement was essentially an *environmental* protest movement. Its struggle for health and safety at work was a struggle for quality of environment at the point of production. Its struggles for decent wages were struggles for environment in the social sphere of reproduction (of the workforce). Few other campaigns have had such impact on the quality of people's environment and lives in the West as these.

Mainstream greens often recoil from this interpretation, but its continuing validity is reinforced by present-day analyses of what 'environmental' move-ments in the third world are about. In India, Kenya and Mexico, for example, they are a 'livelihood struggle' seeking 'to define the benefits of development in terms of basic environmental requirements for energy, water, food and shelter'. And they incorporate 'conservation objectives' *only* in this context – of basic needs (Redclift 1987, 170).

Some greens are also uncomfortable with a second, dialetical, facet of Marxism, which is that it constantly encourages us to be 'historical': that is, to see the state of the world – including nature and our relationship to it – not as fixed or unchanging but related to the specific cultural and economic features of specific societies in particular times and places. This also implies that Marx's writings *themselves* cannot be divorced from the spirit of the times in which he wrote. A corollary is that Marx would write *different* things today, but *without* his overall approach or message being invalidated. Were Marx and Engels alive today, says Parsons (ibid., 69), 'we may reasonably suppose that they would have declared themselves more vigorously and explicitly on the ecological side of their man–nature dialectics'. Grundmann (1991, 80–2) correctly notes that Marx would *not* have been an ecocentric: he ridiculed all forms of nature worship and sentimentalisation. But he agrees that Marx's relative priorities, which were more concerned about wasting human life and labour than wasting non-human nature, did not stem from over-anthropo-centrism: they merely came as a reaction to the most pressing problems of the time. Marx did see nature's value as instrumental' to humans, but to him instrumental value did not mean merely economic or material. It included nature as a source of aesthetic, scientific and moral value.

What critics often find maddening is that the accuracy or otherwise of 'predictions' in Marxism is not the petard by which it might or might not be hoisted. Marxism quite legitimately presents a moving target whereby 'new knowledge and practices mean always, as Engels said, that "materialism must change its form"' (Parsons 1977, 28–9). By its *very definition* dialectical knowledge must be continuously informed and brought up to date.

Indeed, some of Marx's 'laws of history', especially concerning the supposed pivotal role of the proletariat in effecting social change, appear to have been proved incorrect – at least so far (see Chapter 3.7). But this does not mean that Bookchin (1980, 301) is justified in branding Marxist *method* itself as having experienced 'a century of failure, treachery and misadventure'. In fact the method offers ecocentrics two further immensely useful perspectives.

One is upon social change to a radically different society – not so much in predicting how precisely this will occur, but in reminding us not to neglect the importance of changing the *material* organisation of society if we are to achieve it, and, furthermore, reminding us that it *is* achievable. Collectively we *can* shape our future and 'make our own history': an ecological future if we like.

The other perspective is that afforded on capitalism: today's globally pervasive economic system. Without understanding this system we cannot understand why it interferes with environmental systems 'to such an extent that this threatens [our] continued existence. Only when this analysis is available' can we consider exactly how the environment might be protected from the undesirable consequences of human use (Johnston 1989). It is Marxism's *socioanalysis* of capitalism which makes it so penetrating, revealing and ultimately 'shocking', as Heilbroner (1980) puts it. This is an insight into the 'social relations' that underlie the economic system's features and laws ('social' relations here includes relations between people, and between people and nature). These relations are *not immediately clear* to us – for instance we may see a commodity as a 'thing' rather than the set of social–economic relationships between people that it really is a product of. So a level of (social) reality beneath the surface of history is revealed. Marxism is therefore a structuralist approach to history and to society–nature relations.

There are so many Marxisms and Marxists that it is difficult for us to perceive that they have any coherence. This is why Heilbroner's (pp. 20–1) definition of the four essential elements of Marxism is useful. If your perspective on history and social change embraces them all, you are a Marxist. A view of capital which starts from Marx's *socioanalysis* is one. A second is a *dialectical* approach to knowledge, considering the 'innermost nature of things to be dynamic and conflictual rather than inert and static'. A third is its *materialist* approach to history – highlighting the importance of how we organise together and relate, socially, to produce the means of existence, and how this influences other aspects of society, producing class struggle as an element of social change. The fourth, following on the belief in not merely studying the world but in changing it (praxis) is the commitment to *socialism*.

It is worth noting here that some Marxists distinguish between the terms 'socialism' and 'communism' as historical terms – the latter being the ultimate and most desirable stage of human freedom which is to come, while socialism is an intermediate stage (the former Soviet Union may or may not be regarded as an example). Others, however, regard the terms as synonymous and interchangeable: both describing an ideal state that has not yet been anywhere attained.

Within these broad parameters we can distinguish several schools of Marxist thought. Eckersley (1992, 9–116) defines (i) orthodox' and (ii) 'humanist' schools, based respectively on Marx's later and earlier original works, (iii) a humanist neo-Marxism of the Frankfurt School of critical theory (Horkheimer, Adorno and Marcuse all built on and changed aspects of Marx, and Habermas carried this project further and transformed it), (iv) eco-socialism: a 'post-Marxist synthesis'. She considers that orthodox Marxism emphasises how historical progress depends on freeing ourselves from nature by subjugating it via production and technology. Humanist Marxism, by contrast, wants to reassess Marx's technological optimism and belief in material progress, but it still rests on ideas of anthropocentrism and controlling nature. Critical theory also broke away from the so-called scientism of orthodox Marxism's historical materialism, which was alleged to have proposed cast-iron historical laws of historical progress based on economic development. Critical theory worried about the domination and exploitation of people, (and of nature: exploitation that was as much, it judged, a function of cultural ideas and attitudes as of economics. It felt that the development of productive forces alone does not provide true freedom, but instead might lead to a domination and alienation of humans *and* nature. Spreading values of instrumental rationality (involving the drive to efficiency and only valuing things according to narrow economic use) increasingly dominates the lifeworld of humans, and their environment. But critical theory was concerned to 'rebalance' such rationality with concern for feelings, emotions and aesthetics; economic to be balanced with non-economic, cultural values; and materialism with idealism. It fed many ideas, via Marcuse particularly, into 1960s environmental and other radical movements. Critical theory also contributes much to the perspective of eco-socialism as interpreted by Eckersley (ibid., 116–32), though less to the ecosocialism proposed here in Chapter 5.

Carter (1988, 6) analyses Marxism somewhat differently. First, there are 'economistic' Marxists, seeing economics as the main driving force behind society, history and social change. Second, technological determinists give primacy to the 'productive forces' (especially technology) as the device which allows for greater competition within capitalism and at the same time enables capital to control labour in various ways. Third, interactionists' see things in terms of more balanced and dialectical (see below) relations between the various influences on society, and, fourth, there are those who emphasise class struggle as the central dynamic of social change. 'Orthodox' Marxists generally belong to the first and fourth, and perhaps second, of these schools, humanist and 'neo-Marxists' to the third. Concomitantly there tend to be two types of response to ecological problems, as we have noted. Humanist/neo-Marxism sees the Promethean attitude to nature in Marxism as indefensible. But orthodox Marxism connects ecological crisis firmly to capitalist productive relations, which exploit nature in the same way as they exploit people. Hence orthodoxy argues that it is not Prometheanism which is wrong, but capitalism, which prevents social and economic development from being benign to nature (Grundmann 1991, 50).

In what follows, distinctions are sometimes drawn between orthodox and neo-Marxisms where this is considered to be helpful. But this division is not overemphasised, for two reasons. First, many 'orthodox' Marxists would refute such characterisations of them as those of Eckersley or Carter: orthodox interpretations of Marx, they claim, *incorporate* humanism and do not propose nature's subjugation, and also attach considerable importance to culture, ideas, etc. Furthermore, some humanist/neo-Marxism is so idealistic, so prone to reject class analysis and so on, that it effectively is not Marxism at all. Second, as I have said in Chapter 1, the point of this book is not to save Marx or establish authentic Marxism. Rather, it seeks by examining what

Marxists and anarchists have written, to define a radical and coherent eco-socialism, hence it can be eclectic about what Marxisms and other left ideologies it draws on.

3.2 THE MATERIALIST APPROACH TO HISTORY

Base and superstructure

Chapter 1 has already outlined the difference between materialist and idealist approaches to social change, and the importance of this issue to ecocentrism. Marx's approach to explaining how societies have and may evolve is fundamentally materialist.

This does *not* mean that Marx was an 'ontological' (to do with the nature of being) or 'epistemological' (to do with the nature of knowing) materialist. That is to say, he did not assert, as some greens mistakenly think he did, that the world is entirely composed of matter, or that what is known about it can exist 'as an objective material reality totally independently of the subject' (i.e. the person who is seeking to know the world) (D'Arcy 1970). It *does* mean that Marx rejected the notion of history as simply the progress of ideas: the march of a universal human consciousness or spirit which has virtually independent existence from material life. Hence, in making *material* life the starting point of this conception of history, Marx rejected Hegel's idealistic vision where historical progression to freedom meant the realisation of the *'idea'* – a growth within society towards pure rationality and the ideals of freedom and reason that underlay the French revolution.

Marxism, however, does see or hope for history as a progression: to socialism and ultimately communism. (As I have said, for some Marxists, such as the SPGB (Socialist Party of Great Britain) the two terms are synonymous: for others, such as Grundmann, they are not). It is, however, not an inevitable progression, although problems which give rise to people thinking about socialism are inevitable (D'Arcy and Baritrop 1975). The materialist conception of history starts with the premiss that material *production* and the exchange of products constitute the basis of all society. This, the way production is organised – the 'mode of production' – is important to ecocentrics, because producing things constitutes a way of *interacting with nature*. In production we change nature, as 'raw materials', into socially more useful forms, such as oil into plastics or fuel. We use the 'forces of production', which include labour power and the 'means of production' (raw materials, i.e. nature), and the 'instruments of production' (i.e. technology). We also interact with each other in order to produce things: we organise among ourselves – as individuals we cannot produce a plastic bowl from some oil but socially, collectively, we can.

So the way we relate to nature and each other is strongly influenced by the way we organise production – the basis of our material life on earth. A capitalist mode of production implies 'capitalistic' relations with nature and with each other – capitalist *relations of production*, to which correspond particular political and legal arrangements and particular 'forms of social consciousness'. A feudal or socialist mode of production might imply different relations with nature or between people. This immediately implies that if we want to change social and society–nature relationships we must seek such changes not simply in the minds of people – their insights or philosophies, i.e. their 'forms of social consciousness' – but also in their material, economic, life. This is to say that however much ideas act on and shape the material world we create, they must in the first instance be produced *within it*, and must in some general sense be compatible with it (Heilbroner 1980, 63), or they will not find favour as conventional wisdom, i.e. 'common sense'. (What is 'common sense', it follows, is *not* common to all cultures, but varies with different world views, corresponding to specific modes of production in specific historical periods.)

> In the social production which men carry on they enter into definite relations that are indispensable and independent of their will. . . . The sum total of these relations of production constitutes the economic structure of society – the real foundation on which rise legal and political superstructures and to which correspond definite forms of social consciousness. The mode of production in material life determines the general character of the social, political and spiritual processes of life. It is not the consciousness of men that determines their existence, but, on the contrary, their social existence determines their consciousness.
>
> (Marx 1859)

Heilbroner (1980, 87) remarks how 'the Cartesian–Newtonian world view of the seventeenth and eighteenth centuries *sprang from and supported* a society of expanding and prosperous commerce and banking' (emphasis added). In this way Marxists do not see the processes through which ideas and values are accepted into or rejected from conventional wisdom as just fortuitous, or merely governed by rationality or 'common sense' in a vacuum. Rather, to put it simply, Marxists encourage us to see a relationship – a *correspondence* – between the economic mode of production, i.e. what happens at the material *'base'* of society and the prevailing values, morals and ideas, and their enshrinement in the institutions of society (government, law, education), i.e. the *'superstructure'* of society.

For example it is not coincidental that in capitalist societies 'freedom' of the individual is championed, and interpreted particularly as the freedom to own land and other resources, to go into business with minimal planning and taxation restrictions from the state, to compete and to buy and sell what one likes if one can afford it. But it does not include freedom, as of right, from material want or from unemployment. Such a definition of freedom, as part of 'common sense', is important to the functioning of the system. For if, for example, it were people's common sense' view that freedom of course meant freedom to work for a wage if one wanted to then the existence of high unemployment might cause more dissent and unrest than it does.

From this correspondence between base and superstructure it follows that superstructural changes – including radical changes in ideas and values – are unlikely to come about very quickly or coherently without corresponding changes in the base – i.e. the economic, material mode of production (with its corresponding set of social and society–nature relations). The would-be radical, rather than preaching different ideas and lifestyles *in vacuo*, must therefore more profitably attempt simultaneously to change the material context to one in which those ideas and lifestyles can operate relatively unimpeded. As things stand, the relations of production are reinforced by institutional arrangements (laws and customs), and the dominant ideas in society (those which go to make up common values) tend to be those which are conducive to maintaining the dominant class. Such ideas include:

- Belief in the virtues and 'naturalness' of hierarchy, of competition, of struggle for survival and survival of the fittest.
- The Protestant work ethic.
- Equating progress largely with material advancement.
- Equating individual success and fulfilment with material consumption.
- The belief that if you are poor or miserable in this life, there is compensation in 'after-life'.
- The belief that values, emotions, intuitions, spirituality are all secondary to 'hard' economic 'facts of life'.
- The belief that economic laws have the status of 'natural' physical laws.
- The belief that the nuclear family unit is the most desirable form of social organisation between people.

These ideas are all in fact *ideologies*. This means, in the Marxist sense, that they contain unchallenged assumptions which support class interests: i.e. of the 'bourgeoisie' – the elite who own the means of production and decide on their use, and manage the processes whereby economic and social production and reproduction (perpetuating the system) are carried out. They do not come from any 'objective' assessment of the way nature and society 'really are'.

In all this it is not supposed that 'evil' top-hatted cigar-smoking capitalists sit round conspiring to shape the superstructure of ideas to their own advantage: rather that it is an inevitable structural feature of any class society that such deformations should happen. Exploited and exploiters are subjected to equal 'brainwashing'. The corollary to socialists is that they should *not* occur in a classless society.

For greens the implication of the base-superstructure analysis is that it is doubtful whether any society founded on their favoured ecocentric values of spirituality (rather than defining our being via materialism and associated consumerism), cooperation (rather than competition), subjectivity (rather than approaching knowledge of the world as if it existed as a separate 'objective' reality) and the emotions (rather than rationalism) could survive long as a *capitalist* economy. Ecocentric values are therefore often *inherently anti-capitalist*.

Unfolding history

History seen thus, in material terms, focuses on the succession of different modes of production, and all other changes are seen in the light of this succession. Grundmann (1991, 212) illustrates this by interpreting how that part of Marx's model which says 'forces of production determine mode of production, which determine relations of production', applies historically through different modes of production from (1) antiquity to (2) feudalism to (3) capitalism to (4) (future) communism. The main technologies – forces of production – correspondingly evolve from (1) tools to (2) tools and manufacture to (3) manufacture and machine to (4) (uncertain). The forms of production evolve through (1) production for use value (2) for use and exchange value (3) for exchange value (4) back to use value. The purpose of production evolves as (1) for subsistence needs (2) for subsistence needs (3) to generate surpluses (4) for needs, and the form of socialisation as (1) social production (communally) regulated by blind rules (2) political regulation through guilds and estates, with some markets (3) market regulation of independent producers (4) social production regulated by conscious plan.

Parsons (1977, 3–4) presents a sequence highlighting how each mode of production has a specific manner of relating to *nature* – changing Western conceptions of nature historically were compatible with specific modes of production as follows:

> *Neolithic mode of production*: nature as a mother, sacred and with power to dispense good and evil to the 'child', which is society.
>
> *Slave-owning*: nature as a supernatural despot arbitrarily disposing of people and things of the lower orders and rewarding and punishing servants.
>
> *Feudal*: nature as a compact hierarchical chain of being, where each link is interdependent; organically changing but maintaining the hierarchy.
>
> *Capitalist*: nature as an atomic mechanistic system devoid of innate value, purpose and spirit – its value being controlled by the laws of exchange.

It might be added that in early modes of production what humans could do was often circumscribed substantially by natural conditions. Later on, however, nature was increasingly changed in form, so that ultimately, in capitalism, nature could be said to be entirely *produced*, materially and conceptually, by society (see Chapter 3.6).

Parsons detects a new, 'fastest growing' perception today – a post-capitalist' one. This is dialectical (Chapter 3.6) and it will eventually fit with the coming socialist mode of production, where the working classes have taken over the state to operate society in the interests of all and, ultimately with communism, a classless society with no need for a state.

Smith's (1984) account of Marx's historical materialism stresses how production and reproduction of material life are done *socially*. Initially it is founded on a simple division of labour between the sexes (producing different consciousness between them) where society and nature are harmoniously balanced and consumption matches production. The value of products comes only through their *use* for people.

But because of the possibility of lean years, people desire to produce and carry over surpluses. These eventually become a feature of society, allowing further division of labour (because some production can be for exchange, not just subsistence) and requiring specific forms of social organisation. This is the basis for a *class*-divided society. In a slave economy a class has appeared which appropriates the surplus but does little productive labour itself. For this class the relation with nature becomes more indirect, and mediated by institutions that maintain surpluses and appropriate social relations.

In the further changes of the economy to one where production is (at least partly) for the purpose of exchanging the use value of one product with the use values of others, there are further changes of social relations and institutions. Markets appear, to centralise, simplify (in theory) and regulate exchange (encouraging the growth of towns). And with the appearance of a class dedicated to accumulating surplus value (see Chapter 3.3), production eventually becomes geared to this purpose. Nature therefore is converted into *commodities* (in which there is exchange value, simultaneously with use values).

The change from feudalism to capitalist modes of production particularly interested Marx. Johnston (1989, 44–50) characterises feudalism as a *rank redistribution* society (wealth and power distributed according to people's rank), as opposed to preceding primitive communist societies, based on *reciprocity*, common ownership and democracy of a kind. Fundamental to feudalism was the uneven distribution of power: the twin bases of power being land ownership and the institution of serfdom. The most fertile land was not held in common, but individually owned. Serfs were peasants who had to work partly for a landlord – paying tithes in money or other forms of due. The serfs were not free to work for whom they liked: neither were apprentices of craftsmen in towns. Products were not generally traded on open markets; they were either directly consumed or transferred to the elite classes. However, the increasingly grandiose lifestyles and pretensions of the ruling class did encourage them to some trade – especially for the sought-after products of the East. This, plus the pressure on peasants to produce more (in order to pay increasing taxes or to earn some money in trading for themselves) constituted a growth dynamic. This again differentiated feudalism from primitive communism, rendering the former ultimately in disequilibrium with its environment, unlike the latter.

Despite this, feudalism is generally seen as a mode of production where people were close to the land and working with, rather than against, its cycles. (This prevailing wisdom is challenged: Cooter (1978), for instance, draws attention to the many ecologically damaging aspects of medieval open-field agriculture.) They were tied to the land, in a way, as they were tied to each other. The relations of production were multifaceted. Though peasants and apprentices were economically bound to landlords and craftsmen, the latter were obliged to look after the former's spiritual, moral and material wellbeing. It was a close-knit *gemeinschaft* society; oppressive in many ways, but 'ecocentric' in some ways too. For instance:

> there was also much hard, boring work, but it was performed in the company of others, in a leisurely tempo which allowed chatting and singing, allowed playful relaxation, homely tenderness and aesthetic ideas, 'This economy offered something of importance which we have almost forgotten today, namely leisure: not freedom *from*, but freedom *in* work' [Mumford 1934, 56]. Of course, 'One must be very cautious and not idealise the pre-industrial handicraft economy'. It had also its negative aspects, but one should be cautious in calling it primitive and unproductive. 'In overall balance, many of these pre-industrial technologies were . . . "productiver" and more appropriate to man and nature than our substitutes of today'. . .
>
> (Sarkar 1983, 56, citing Ullrich 1979)

Marx's accounts of the transition to capitalism make much of the removal of people from the land as a removal, and alienation, from *nature*, which is of much consequence to ecocentrics. Not only was feudalism replaced by capitalist agriculture which damages the long-term fertility of the soil in search of short-term profits, but also a state of mind was created in which people no longer appreciated the connections between the land and what they consumed every day, and did not see the countryside as a place of production and power relations, preferring to regard it through romantic lenses as an idyllic place (see Short 1991, chapter 4).

Historical and economic determinism?

Greens are among many critics who declare that Marx's view of society and history is deterministic. That is, it is accused of asserting that everything non-economic about a society is shaped by economics (crude economic determinism), and/or that there is some inexorable pattern of development through which societies must evolve (amounting to phases, predictable through scientific 'laws' of history – i.e. historical determinism).

Orthodox Marxists sometimes come close to affirming this view. On the other hand, neo-Marxists influenced by, or part of, the 'Frankfurt School' distance themselves from it – sometimes so much so that they may abandon Marxism's fundamental tenets.

D'Arcy (1970) argues for orthodoxy. Productive forces at the base, he says, *determine* relations of production, which *determine* the ideological superstructure. This does not, he continues, indicate that there is no place for an independent consciousness or a political culture, rather that productive forces, relations and superstructure are linked in a system of interactions governed by this broad materialist principle. And neo-Marxist interpretations, that see forces of production, including technology, changing in response to changes in class relations, and advocate 'Hegalian Marxism' (see below) are 'quite contrary' to Marx – and to the SPGB, on whose platform D'Arcy speaks.

But Heilbroner points out that neither the forces of production (the population, its skills, arts, techniques and artefacts) nor the production relations are narrow economic concepts. While all their elements are organised around production, this does not mean that economic motives are thought to dominate all others. The 'production and reproduction of real life' (Engels) may be an *ultimate* determinant of history, but not the only one. Thus the crude economism which has occasionally marked Marxist historiography is not, says Heilbroner (1980, 66), *inherent* in its materialist emphasis. For historical materialism has a *dialectical* element, which implies several things.

First, whereas determinism suggests one-way cause–effect relationships between discrete elements (*a* causes *b*, which then causes *c* – as in classical science), the dialectical view does not see economic base and ideological/cultural superstructure as separate from each other – they are part of, and inherent in, each other, incapable of existing or being defined separately. And they constantly interact to change each other. Thus our originally simple base–superstructure model is really far more complex than one which suggest that the former rigidly determines what goes on in the latter.

Second, historical *change* also occurs through a dialectical process. At any given time in a society's history, because society is not stable, there will be tensions between the existing state of affairs (the existing social and economic arrangements, for example) and groups to whom that state of affairs is unsatisfactory. Through a process of struggle and conflict between the old (the 'thesis') and the emerging order (the 'antithesis') a new form of society is reached – a synthesis between old and new. The contradiction between thesis and antithesis has thus produced a 'synthesis'. This can itself be seen as a new thesis, which is not stable but also contains inherent contradictory (opposite) elements. Through the interaction of these new opposites further change will occur. This is the dialectical process in history – a process of dialogue and interaction between opposite, contradictory elements impelling us through social change hopefully to an ultimate state – communism – where there will not be class conflict.

Thus the change from feudalism to capitalism arose as a class of people developed who wanted to employ (buy) labour. But labour was not free to be bought and sold: under feudalism much of it was tied. So began a struggle between the feudal aristocracy and the new entrepreneurial class of merchants and manufacturers. The existing state of affairs was progressively negated by the growth of its opposite, and the seeds of the new system (capitalism) grew out of the old.

As D'Arcy and Baritrop (1975) phrase it, at a particular stage of development of the forces of production they came into conflict with the old social relations, which become a brake on their further development. New groups who wanted to develop these forces (e.g. capitalist entrepreneurs in feudal society, or the proletariat through collective associations in capitalist society) were prevented from so doing because they did not have access to these forces (e.g. land, labour and technologies were owned by the old dominant groups – feudal landowners in feudalism, the bourgeoisie in capitalism). Hence the new order challenged the old, and eventually developed a new mode of production with new social relations. However, this process was modified, or conditioned, by the fact that a given society cannot be fully broken up until its productive forces have been as fully developed as is possible, and until that society has produced the conditions which make its own demise possible (e.g. the emergence of a new class, with its own class identity and consciousness). Hence proper communism could not come out of the Russian revolution because, although the *ideology* of socialism existed with the Bolsheviks, the productive forces had not been developed sufficiently under feudalism to provide for all the features of a socialist state, including socialist relations of production. A phase of capitalist development (say some Marxists), with its prodigious development of industry and productive forces, is always necessary to bring people the material wellbeing that they need before their energies will turn to setting up a socially just and non-materially more fulfilling society. At first the Russian revolution appeared to contradict this idea through promising to provide a transition directly from feudalism to 'communism', and it set Marxists saying that a capitalist phase might not after all be strictly necessary. But in fact a form of capitalism – *state* capitalism was indeed set up. And it may well be succeeded in the 1990s by versions of the 'free' market, so that Russia will not have avoided a capitalist phase. If this model is correct, of course, a further implication is that people will not create an *ecologically* sound society – which is a post-capitalist one – until they are materially reasonably provided for.

All this sounds as if general principles or 'scientific laws' of 'history *are* indeed being expounded in this theory of history. However, Heilbroner points out that the 'scientific' approach to history spoken of by Marxism is different from the approach of positivist science (the kind of science that has grown dominant in the West, and prioritises so-called 'objective facts'). Whereas the latter is a method of obtaining information about the world by formulating refutable hypotheses, to Marxism 'scientific' truth means something rather different. It simply means the *real* truth: that is the *underlying* essence of how society operates as distinct from surface appearances. 'Scientific laws of history' means what is *really* happening in the substructure of society – as revealed by socio-economic analysis, materialist history and dialectical reasoning – rather than just what presents itself superficially.

Heilbroner further maintains that dialectics, of themselves, while identifying forces important in making history, do not *predict* (which is what positivist science sets out to do) any specific resolution of history, as some critics (e.g. Carter 1988) wrongly believe they do. Other Marxists make a similar point to Heilbroner's:

> Marx's dialectic is a quite different mode of thought from the formal logic of positivist science, and it is symptomatic of the latter that its adherents often cannot conceive of different logics.
>
> (Smith and O'Keefe 1980, 32)

And Williams (1983) affirms that we are not dealing with 'categorical laws of regular determination' operating uniformly through space and time, but 'historically specific determinations' which do not hold good in all circumstances.

But if there seems to be a suspicion of fudge here, it is strengthened by Marxism's utopianism: the vision of a free, classless, unalienated future communism where people will control their destiny – make their own history – in true consciousness of their own power and freedom. For that vision, if not strictly inevitable, does seem a likely outcome of history when viewed from Marxism's perspective.

As Heilbroner puts it (1980, 37), Marxists find uncomfortable the idea of a never-ending struggle without transcendence, hence their view of history

> is not content to declare that ceaseless change is inherent in history as an aspect of the nature of all reality. It imposes a design on the course of history . . . in no wise less idealist than the vast mystical resolution attributed to history by Hegel.

Neo-Marxism

Neo-Marxists have been uneasy about over-determinism and millenarianism (i.e. believing in the millennium – the coming of salvation after a long historical period). Habermas considers that despite its original project, Marxism has contributed to an advance of positivism (seeing history as a deterministic unfolding of laws and principles) by stressing too much the purely instrumental dimensions of labour and production (Kearney 1986, 224). Marx's original project, for Habermas, was meant to synthesise critical reflection (ideas) with praxis (bringing about material change), resolving the opposition between idealism and materialism. In other words it was meant to gain knowledge of processes in history not in order just to comment passively and objectively on their unfolding, but to be critical about the form of society emerging, and to act on it to change it to a society judged to be morally better, i.e. to shape the propitious conditions for a socialist future.

Lukács and Marcuse were among those neo-Marxists who revised Marxism to make it bring about such a resolution. Lukács 'injected a strong dose of left-wing Hegelian dialectics back into Marxist theory – then in danger of becoming a positivist system akin to that of natural science', says Kearney (ibid., 138). He stood against the determinist orthodoxies of official 'vulgar' Marxism (Stalinism). Its economic reductionism, where everything was adjudged to be the outcome of economic forces and structures, ignored the creative role of human consciousness and decisions (ideas), being able to act independently of economics. Thus Stalinism deformed Marx's original perspective on the dynamic relations between theory and practice. Lukács's view that 'without the necessary renewal of theory [in this direction] there can be none of practice' became a keystone of the New Left movement (ibid., 139), which fed into 1960s environmentalism.

Lukács rejected both Hegel's idealist emphasis on the autonomous powers of consciousness, and Engels's position, which gave the 'subject' no power at all to change things at will. Engels's interpretations, thought Lukács, suggested a rigid historical causality, assuming that reality unfolds according to determinist laws independent of human initiative. Instead, Lukács re-stressed the concept of a 'dialectical rapport' between consciousness and material reality (i.e. between idealism and materialism), the individual and society, base and superstructure – and society and nature. Atkinson (1991) and Grundmann (1991) are two Marxist-influenced ecological writers who seem to favour this position. The latter (p. 213) reformulates the three-level model – base (mode and relations of production and productive forces) determines or conditions superstructure (politics, law, etc.) which determines or conditions culture (forms of social consciousness) – to argue for the autonomy of each level, and of the political, economic and scientific 'subsystems' in society. Hence the economy can change to reflect changes in law, or new scientific inventions, while politics can resist or enhance certain technologies, and so on. He therefore posits 'functional links' in Marx's model but allows the subsystems to be defined and to work independently. He justifies this (p. 160) by claiming that Marx himself did not precisely differentiate between legal, political and economic factors, and that relations of production and productive forces cannot even be *defined* independently of one another (this being a general character-istic of phenomena which are related dialectically). And, contrary to orthodox Marxism, 'there is no economic alienation from which all other forms [of alienation] are derived' (p. 164).

Marcuse, an inspiration behind the New Left and 1970s ecocentrism alike, also refuted vulgar materialism, that is, economic determinism. He argued that consciousness could transcend the material social conditions which alienated it, to liberate *itself*. Thus people might have to be suppressed and exploited by economic structures before they wanted liberation, but more was necessary – a conscious desire for and ideas about liberation – before it could be achieved. A 'properly *philosophic* appreciation of the essential structures of concrete individual experience' (Kearney 1986, 206) was most fundamental, shifting the emphasis towards reform of the individual self. And the problem about *class* consciousness, which Marx sees as instrumental in achieving liberation, was that it did not admit that each *individual* had also got a potential for radical dissent. Marcuse's position on social change is echoed by greens and other new social movements.

Marcuse did not see himself as departing from Marx; rather as reverting to Marx's early (1844 *Paris Manuscripts*) view of history as the dialectical interchange of spirit or '*geist*' (ideas) and the material conditions of reality. Thus he veered towards the 'Hegelian Marxism' spurned by the more orthodox SPGB. But not all Marxists went in this direction. Althusser vigorously rejected it, and the idea of the primacy of the subject. Human will, creativity, individuality and 'responsibility' he dismissed as bourgeois liberalism (Kearney 1986, 304). The real subject, he maintained, which Marxist analysis revealed, was not human individuals, but the hidden economic relations of production. They ultimately determine what function individuals fulfil in the production system. Althusser thus wanted to emphasise the primacy of this economic structure over the subject.

Hence the debate which Marx started has carried on strongly in neo-Marxism. Reflecting Marcuse's perspective, much of the New Left 1960s radicalism floundered, however, on the sterility of the idealistic personal-is-political/'all-you-need-is-love' approach: the movements's revolutionary potential was quickly dissipated and assimilated into conventional society. Ecocentrism, with its New Age tendency, may be headed in the same direction and it perhaps needs an injection of Althusser's materialism as an antidote. At the same time it needs to be sensible to the ideas of another neo-Marxist, Gramsci. For in his refutation of objective laws of history and economic determinism, and in his argument for the active creative role of human consciousness, he drew attention to the importance of *mass* self-consciousness – the consciousness of groups and whole societies – which have an ultimate role in revolutionary struggle.

The fact that base and superstructure are dialectically and flexibly related meant, he said, that there is an important struggle to be fought; not just for control of the means of production, but for the control of the ideological superstructure – for hegemony over mass consciousness. This struggle has gained hugely in significance with the advent of mass media. For Gramsci, ideological control over mass consciousness, which the ruling classes now exercise, was a decisive a factor in their domination, achieved through legal and military coercion. For it means control over how society sees itself – the conventional wisdom, the system of myths, images and morality that people

77

identify with publicly and privately; the general ethos of the national community that pervades the churches, schools and the family, and the mass media. Greens, like socialists before them, have learnt how difficult it is to shake this total cultural control, or 'hegemony', within the existing relations of production.

3.3 THE ANALYSIS OF CAPITALISM

What is it?

In his highly intelligible summary of capitalism's characteristics, Johnston (1989, chapter 3) points out that while industrialism and market-regulated exchange are significant features of this system, they are not *defining* features. Citing Desai (1983, 65–6), he lists capitalism's main aspects. They are: production for sale rather than direct consumption; buying and selling labour power; exchange through the medium of money; capitalists and their agents determine what is made and how; they control the financial decisions that affect the majority without power to influence these decisions; they compete with each other for labour, materials and markets. An important aspect of this last is a constant drive to produce ever more 'efficiently', meaning to maximise income from sales while minimising production costs – the difference between the two is *surplus value*, or profit.

By *definition* (not because of 'greed' or any other sin) capitalists must accumulate wealth from production (profits) and re-invest it to generate more capital. Furthermore 'capitalism grew out of societies based on inequality and developed that inequality' (Johnston 1989, 52) – and, again virtually by definition, must continue to do so. Early capital was not generated specifically through production, but by entrepreneurs buying and selling the surpluses from the feudal subsistence economy and the fruits of overseas exploration and colonisation. This is known as merchant capitalism.

The succeeding phase, of industrial capitalism, dates from the eighteenth century. In it, the forces of production are physically transformed into commodities – goods and services produced not just for their usefulness but in order to be exchanged with other commodities and/or for money. In unfettered capitalism commodities are produced *primarily* for the purpose of generating profits – their social usefulness is secondary. Production will not generally be undertaken unless the market indicates that there is an exchange value to be realised, which is greater than the cost of production, i.e. the commodity can be sold for a profit.

To facilitate production and circulation, money capital has to be converted into forces of production, including labour power (productive capital), and it has to be moved around between commercial transactions. This is the sphere of finance capitalists, who hold, exchange, borrow and lend money. By lending money to industrialists they reappropriate (through interest) some of the surplus value which the industrialists have appropriated from the labour force which produced the commodities.

Commentators often distinguish between a 'Fordist' phase of industrial capitalism, dominated by large-scale, production-line techniques making standardised goods for a mass market, and a more recent 'post-industrial' phase of flexible accumulation. Harvey (1990, 173–9) reviews the transition from Fordism to flexible accumulation, which he sees as a way of coping with the contradiction of overproduction (see below) when more traditional ways of coping with it have been exhausted. It involves flexible, small-batch production for individualised consumption and more 'flexible' (i.e. less secure) employment patterns. There is also (i) spatial integration of production rather than spatial division of labour, (ii) decentralised production with (iii) local negotiations on employment conditions (weakening unions) and (iv) withdrawal of the state from regulation. The first two of these are consonant with green libertarianism, the last two are not: they encourage poorer work environments and greater environmental impact.

Capital itself is succinctly defined by Johnston (1989, 51–2). It is

the result of labour, specifically the surplus derived from employing labour. That surplus is created because the income from the sale of the products of the purchased labour is more than the costs of employing it, and the land resources on which it worked.

Labour and value

The ability to generate capital in production, then, is based on access to land (nature's materials) and labour power. While, under feudalism, labourers were not free to work for anyone, in capitalism they are. Their labour has therefore itself become a commodity to be bought and sold, like any other commodity, as labour power.

In the controversial labour theory of value – the AL theory described briefly in Chapter 2.2 – Marx argues that a commodity's ultimate value does not derive from its availability in relation to the demand for it, but from the labour invested to make it more socially useful. To see this, consider a situation where supply and demand for a commodity were exactly balanced: one must, then, look elsewhere for a source of ultimate value common to all commodities. Labour is that common source.

This does not deny that an element of value also comes from nature's raw materials, and, by implication, that nature puts some ultimate limits on wealth. Indeed, the term 'means of production' includes nature's materials, so use value is derived from a combination of them, and human labour. However, nature's materials are seldom useful *until* they are converted to useful form by labour. But then, since Marxism regards humans as a part of nature, labour represents nothing more than nature working on *itself* to change its form.

Commodity values in capitalism can be expressed in units of 'abstract' labour – a concept whereby different sorts of labour, e.g. mental and physical, can be made equivalent (Harvey 1982, 10–15). The commodity's value, then, as expressed through the medium of money at the time of exchange, is *fundamentally* the amount of abstract labour put in to producing it. In fact the price of a product does not *totally* reflect its value thus defined, for discrepancies between supply and demand can lead to prices above or below the abstract labour value (i.e. the value of the labour in it) (Heilbroner 1980, 118). This is clearly seen when monopolies are able to hold prices 'artificially' high.

In this context Marx distinguished between value as represented by the *labour* put in a commodity, and the value of *labour power*. The latter is what the labourer is actually paid by the capitalist and it is less than the value of the labour which has been invested in producing the commodity. The difference between the two represents profit or surplus value. What the capitalist pays the labourer can also be thought of as the value of labour which needs to be given back to the labourer in order that he or she can afford to live. It is therefore the value of labour which needs to be expended to reproduce the workforce. It is more than just the costs of subsistence, for it has a historical and moral element related to what society considers to be an acceptable standard of living – the 'socially necessary' labour power (Hardy 1970). To repeat, capital is brought into being (fundamentally, and leaving aside fluctuations in supply and demand) when the capitalist pays less for labour power than the value of the labour, as realised in exchange, put into making the products. Capital, then, is the expression of the difference between the two – the surplus value. In other words, labourers are exploited: they get paid only part of the total value of their labour – paid for only part of the true value of their working day. The rest, the surplus value, is creamed off, or appropriated, by capitalists. Their place in the social economic system totally depends on labourers producing surplus value, and capitalists being able to expropriate and accumulate it. They can do this only because they own and control the forces of production (including labour itself, which they buy on the open market).

Smith (1984, 48) emphasises this, to demonstrate how capital accumulation is therefore predicated on a specific *class structure*. Hence what to most of us seems an impersonal economic process of production (the superficial appearance) is really an expression of social (class) relations (the substructural reality). Put at their simplest, these relations, this class structure, is the two-class society described in the *Communist Manifesto*: the *bourgeoisie*, who own and control the means of production, distribution and exchange, and the *proletariat*, who have only their labour to sell. As Heilbroner points out, even when labour power is being well paid, the value of the labour that goes into products and is realised when the products are exchanged must always be greater than what is paid for labour power if the system is to survive. The system depends on class exploitation.

Johnston (1989, 53) illustrates how the two-class model has become more complex with time:

> As capitalism expanded, the capital investment requirements were increasingly greater than individuals could meet from their amassed wealth. . . . It was necessary for them . . . to combine and borrow from others to obtain the necessary sums. Thus the precise distinction between labourers and capitalists disappeared . . . increasingly the difference within capitalism is between the managers of capital and the workers. . . .

Many people think that the bourgeois–proletariat model has been in fact invalidated by the advent of (a) a huge middle class of small investors and (b) increasing difficulty in locating power with obviously identifiable individuals like Ford or Rockefeller. Indeed Gorz (1982, 52) believes that in the post-industrial ('post-Fordist') society capitalists have become mere 'functionaries of capital', just managing a system. And it is *the system itself*, not individuals within it, that is the repository of power. But the third world lumpenproletariat who face starvation each year may not be convinced by this argument. Indeed, it breaks down even in the West where, in the 1980s, disparities between rich and poor actually grew (see Chapter 1), and where it is one thing to own shares in a company but quite another to have any *control* over what it does. Not even national governments have been able to control the machinations of multinationals in the 1980s, or the rashes of takeovers, asset-stripping exercises and the like.

Competition and productivity

It is important to stress that the relations of production under capitalism changed radically from what they were under feudalism. With 'freedom' to buy and sell labour (and to be unemployed), and with the regulation of exchange via the market these relations became, among other things, competitive. Worker competed with worker to sell labour (competition being more acute when there was high unemployment) and bourgeois competed with bourgeois to sell, and to accumulate capital. In an open market, as we have so often heard, capitalist laws of economics decree that competitiveness must be maximised, along with profit margins, by maximising *productivity*. This means maximising the gap between costs – in labour power and overheads – and returns – in exchange value (i.e. largely the value of labour) – per unit of output.

The most potent ways of increasing productivity per worker are through division of labour into specialised tasks and their automation and routinisation via the production line ('Fordism'), applying time-and-motion theory to them ('Taylorism' or 'scientific management') and replacing labour by machines. These processes de-skill production, further decreasing the market value of labour power (wages). Productivity increases in Fordism are also sought by increased scales of production, in 'vertically integrated' units – the Ford plant at Dagenham being a classic example, with pig-iron and coke going in at one end and fully-assembled cars coming out at the other.

As Johnston (1989, 55) says, increased productivity is most necessary for capital accumulation when competition is intense between producers. It is particularly pressing in the declining markets during the periodic recessions which characterise capitalism. In the 1980s 'rationalisation' through sacking labour and intensifying mechanisation (via information technology), together with other measures to drive the real value of labour power down (productivity deals, longer hours, lower wages, part-time jobs, poor conditions) were justified by appeal to 'laws' of supply, demand and competition. Thus social conditions in Europe and North America changed radically in response to (economic) forces perceived to be beyond the control of individuals or groups, firms or governments.

As O'Connor (1988) points out, crises such as recession pconstitute a *necessary* part of capitalism. They allow capitalists to restructure productive forces (e.g. to mechanise and automate). They can also restructure productive relations, destroying union power via the threat of unemployment, and getting control of more production – because the weak firms go to the wall and are taken over – and of markets. And they get maximum help from the state, which improves the physical and financial infrastructure, desperately trying to encourage 'an entrepreneurial climate'. Most people put up with the immiseration of their lives which all this entails (unemployment, de-skilling, impoverished environments) because they believe that 'There is no alternative' to what Margaret Thatcher called 'the laws of economic gravity'.

Habermas thinks that the role of state intervention in all walks of life gets increasingly important in 'late capitalism' when the system shows signs of strain (Kearney 1986). The role of technology is two-fold, On the one hand it is the means whereby firms intensify their competitiveness (getting greater productivity and devising new products to tempt consumers). On the other hand it helps to reduce labour costs, it can work twenty-four hours a day and seven days a week and will not go on strike, and it de-skills jobs, making for cheaper labour power needs. And it creates unemployment – a powerful *weapon* that was used to great effect by Western governments in the 1980s to defeat union power and undermine wages and conditions.

Contradictions of capitalism

Not only are crises such as inflation, depression, the imbalance between production and need, and environmental degradation needed by capitalism, they are also an inevitable outcome of the system:

> The persistence of these failures provides a massive confirmation of the
> fundamental logic of the Marxian analysis . . . [they] flow from the

deepest social and historical properties of the system and not circumstantial accidents.

Here, Heilbroner (1980, 132) is taking a dialectical perspective on capitalism, seeing its *inherent* contradictions – that is, features antagonistic to itself, which will eventually be its downfall. The 'ecological contradiction' is obvious today, and it was the starting point for both Goldsmith *et al.* (1972) and Schumacher (1973), though these non-Marxists never expressed it with Marxian precision – we will examine it below.

It however stems from other contradictions, the net effect of which is to increase even further capitalism's impetus for expansion, and exploitation of labour through appropriating surplus value.

One of the most socially and environmentally exploitative capitalist organisations is the *monopoly*. It arises through a lack of competition, which paradoxically is the logical end result of 'free' competition, when nobody intervenes to save 'inefficient' or unsuccessful producers from going out of business entirely.

Monopolies partly arise because the market does not do what liberal economists like Hayek claim it does: coordinate rationally the actions of economic actors via a price mechanism that communicates all relevant information to them. In reality standard game theory applies to the situation, so that while all producers might benefit from mutual cooperation (informing each other about future production plans to meet demand) those who do *not* cooperate can win more for themselves. This is 'rational' in a way, given firms' *self*-interest, though it does not make for a rational production and distribution system overall. Hence much relevant information is blocked because of competition, and small firms are likely to suffer disproportionately in this situation (O'Neil 1988).

In a second contradiction, capitalism, which needs ever-expanding markets, actually destroys its markets, causing *overproduction*. This tendency can best be envisaged by imagining a closed society or economic system. Here, all the people who sold their labour (producers) would also constitute the major part of the market for their products (consumers). But by definition they cannot be paid enough to consume all the results of their labours: if they were, there would be no surplus value in the system. This problem is worsened by the drive to increase productivity, which lowers wage costs per unit of output, so that demand is further suppressed in relation to supply.

This process is exacerbated by a tendency to a *falling rate of profit*, produced through mechanisation. In larger, more capital-intensive plants unit costs of production will be lower. But as labour is steadily replaced by machines there are proportionally higher depreciation charges and auxiliary materials costs (fuel, raw materials, etc.) to meet. Because of this, under given technical conditions (without further innovation) more investment in machinery will reduce unit production costs further – but in ever-declining proportion to these depreciation and auxiliary materials costs. So each additional (marginal) unit of investment secures a smaller return than the last:

> the more advanced methods tend to achieve a lower unit production cost at the expense of a lower rate of profit. Competition nonetheless forces capitalists to adopt these methods, because the capitalist with lower unit costs can lower his prices and expand at the expense of his competitors – thus offsetting lower rates of profit by means of a larger share of the market.
>
> (Bottomore *et al.* 1983, 159)

Put another way, labour is the ultimate source of surplus value, and more of this value can generally be squeezed out of production, but only at the cost of increasingly expensive capital inputs (if installing a machine enables half the workers to produce the same amount as previously, then productivity has doubled). So the rate at which this process secures increases in surplus value decreases steadily (until, that is, a major technological breakthrough, such as information technology, comes along to lower non-labour costs substantially).

These contradictions can be offset by various means: cheaper costs via cheaper raw materials or expanding demand via more advertising, marketing and product innovation (perhaps persuading people to spend more of their savings), and, very obviously, by spatial expansion of the market and of production. Thus, as Marx and Engels foresaw in the *Communist Manifesto*, capitalism must expand relentlessly across the globe, revolutionising the economies which it penetrates and changing and homogenising different cultures. The opening of a branch of McDonald's in Pushkin Square, Moscow, in 1990 was a potent symbol of this steadily increasing force. 'Hungry Muscovites won't need to queue long for their food, either', ran the advert in *The Times*, 'At McDonald's we aim to serve everyone within three minutes'.

About every fifty years, or less nowadays, crises of overproduction, falling profits and stagnation prompt large-scale withdrawals from production (capitalists transfer their money into more 'fixed' form – buildings, machines, land) thus encouraging monopolies. Restructuring and 'rationalisation' take place, leading, among other things, to development of geographical cores and peripheries within and between nations. Wealth and political power concentrates in the cores, as is appropriated from peripheries. The nature of environmental problems in the latter differs, as remarked earlier, as between these zones. In peripheries such as the third world or the formerly industrial north of England they are often matters of basic survival and struggle against naked exploitation (in sweatshops or next to improperly regulated chemical plants like Bhopal), whereas in cores they are often associated with the spiritual angsts of people who have a lot of material wealth. Alienation is a major element of such middle-class angsts, which is why it figures so prominently in Western ecocentrism. Marxism, too, has much to say about alienation.

Alienation from (human) nature

By alienation, Marxists mean separation or distancing of yourself from (i) the results or products of your own activity, (ii) the rest of nature, (iii) other human beings, (iv) yourself. The first three categories are all considered to be aspects of category (iv), yourself. 'So self-alienation is not just one among the forms of alienation, but the very essence and basic structure of alienation' (Bottomore *et al.* 1983, 10). In this holistic view, what we produce is part of us, as are other people and the rest of 'nature' part of us. So to be separated from these things – these aspects of ourselves – means to be distanced from what is in our own nature. Marxists see our nature, i.e. 'human nature', not as totally fixed or unchangeable, but largely changing through history. Thus not to act according to our nature means not to act in accordance with 'historically created human possibilities . . . especially the human capacity for freedom and creativity' (ibid., 13).

In capitalist society many people (such as the greens; e.g. Roszak 1970) place individual consciousness at the centre of what is frequently regarded, idealistically, as an existential, psychological problem of alienation (this view of alienation is idealistic because it regards the mind as the main locus of its experience). Thus consciousness of being alienated means awareness that 'what I am [my roles in life] is not actually me' (Coleman 1984). It is a disharmony which may be thought not to be self-produced but to result from external pressures over which I have no control.

By contrast alienation in the view of Marxism is produced *by me and my society*, and part of my alienated state derives from a lack of awareness that I could, with others, control things to overcome my alienation. Furthermore, its genesis is material: in capitalist society alienation derives specifically from capitalist relations and processes of production. These are not forces external to me and my community. They are socially induced, in a specific historical period, and so could be socially changed. Thus whereas most psychotherapists do not place alienation in its social context, treating it instead as a problem of the individual, Marxism makes it totally social. And it is grounded in the concept of production as a major defining characteristic of being human. When we produce, we change nature with calculation and forethought, and this more than anything is deemed to make us unlike animals.

In capitalist productive relations the labourer is doing more than selling 'labour' as some kind of abstract thing or commodity – even though it is thought of as such. For, since through labour we produce things, and since in producing these things we change the nature of what we are, then labour is a means of creating what we are – of 'self-creation'. So in selling labour we are selling *ourselves and (parts of) our lives*.

People are bought and sold. So the relationship between them is that between buyers and sellers. It is a cash relationship (channelled through the 'cash nexus' or connection) where the labourers reduce themselves to the status of any other objects for sale in the market place. They are objectified, i.e. reified: reduced to the status of *things*. In this process moral and spiritual matters tend to lose the importance they had in other modes of production, except in so much as they govern the capacity to produce efficiently. Morality becomes subservient to the central question of maximising profit (i.e. maximising capital). Subjective, emotional, spiritual aspects in a relationship often make people do irrational things. Through love of nature or place we might not wish to exploit a particular resource in a particular place, when it would be economically profitable to do so. Through love of people we might be tempted not to sack them even if we cannot 'afford' to employ them without reducing our profits. But in capitalism

> It is possible to manage a business in decline [which requires shedding part of the workforce] and gain satisfaction. The day I forget the pain my actions cause to individuals I am as dead as a dodo. But I can't let emotions override business.

So says the chief executive of VSEL shipyards in Barrow-in-Furness, contemplating the redundancy programme he has steered through the firm (*Independent on Sunday*, 2 August 1992, Business section, p. 12).

So when we turn ourselves into commodities to be bought and sold through the labour market, our worth is measured mainly by our worth in the *market*. Indeed, our everyday vocabulary displays how all aspects of our lives have become perverted to notions of market exchange. We talk of the 'returns' we get from relationships, and whether they pay 'dividends' or are 'profitless'; of our 'stock' with others and our 'assets' of looks or brains; and indeed whether we are in the 'market' for an affair or marriage (Seabrook 1990, 12).

It is not just the 'proletariat' who alienate themselves in capitalism. The bourgeoisie do so too. They do not identify subjectively with each other, valuing each other primarily as people to be cared about regardless of their talents, income, status or expertise. They are in competition with each other to try to monopolise the processes and resources for creating capital. That competition is ruthless and the 'weak' – the less efficient – must go to the wall. There can be no room for sentiment in such an environment; you have to regard relationships 'objectively', and your competitors as objects.

Alienation is also created by capitalist productive processes resulting from the inherent drive to increase productivity, such as mechanisation, division of labour and the production line. They all chop work into bits: small operations to be repeated by machines or by humans acting as 'appendages of a machine'. This process is alienating. Workers are separated from their own creativity, an important aspect of their own (potential) natures, to become 'crippled monstrosities' (Smith 1984, 51). Think of the modern checkout operative, passing goods over an electronic eye all day and compare the skill in this to that of a 'shop assistant' forty years ago, which involved physical and mental agility, an encyclopedic knowledge of stock and prices, and the capacity to empathise with and advise the public. Think of Charlie Chaplin tightening bolts on the production line all day until he goes berserk – no wonder that the film *Modern Times* so displeased Henry Ford and the American capitalist establishment. William Morris's socialism was centrally concerned about this type of alienation: a concern echoed in calls by Schumacher (1980) and other greens for 'meaningful work'.

Marx disapproved of alienation through the objectification of humans under capitalism, via the cash nexus. But in fashioning objects out of nature in socialist production, we can make products solely because they are *useful* or pleasurable to someone. Objectifying nature in this way satisfies the needs of others. This is different from satisfying the created 'wants' of a consumer society. And the products, if they embody the craft or creativity of the maker, therefore reflect the personality of the maker. Hence in fashioning objects out of nature the subject–object distinction between producers (subjects) and consumers (objects of production) can be collapsed. This is all the more true if production can be localised, so that instead of making for an anonymous 'market' we can have some conception of the actual people for whom we are working.

This would be the essence of communal relationships in communism – of human existence as communal beings. Under capitalism, with its liberal philosophy of extreme individualism, this essential *communal* side of human personality – the need to relate to others – is diminished. Hence, people again distance themselves from, and deny, a side of themselves. Here, despite having described humans as ensembles of social relations (in the *Thesis on Feuerbach*) so that 'human nature' must vary with different societies in time and space (with the mode of production), Marx nonetheless clearly propounds a theory of universal human nature: community and creativity are empirically-given universal latent human characteristics (Grundmann 1991, 102).

Labour creates capital, and capital then exploits labour by draining it of social, communal, meaning. If products are made principally for profit, to be realised in a vast anonymous market; if they can no longer be directly identified by producers with the specific needs of specific consumers, then the reverse also applies. As Harvey (1990, 101) graphically puts it, we are hidden from

> The conditions of labour and life, the sense of joy, anger or frustration that lies behind the production of commodities. . . . We can take our daily breakfast without a thought for the myriad people who engaged in its production. All traces of exploitation are obliterated. . . . We cannot tell from contemplation of any object in the supermarket what conditions of labour lay behind its production.

It follows, of course, that we cannot, either, tell anything about the relationship between its producers and nature.

Thus the meaning of the producers' labour in capitalism lies not as an expression of their sociality, but in the money they are paid for doing it. Hence the ludicrous paradox in which we cannot, through our labour, create decent conditions of life for most of our third world brothers and sisters. In fact we have to remedy through charity the *harm* we have done them through our labour, which often exploits them directly or indirectly through its profit orientation.

And labour has allowed a state to be set up by the bourgeoisie, to support, advance and defend their interests. The state at present polices the private property system and the market economy. Its ostensibly impartial system of justice actually protects the interests of capital against the working classes, and it protects all the privileges and trappings of a hierarchical society. The state takes power to itself, away from the people. It is supposed to act for all, but in reality government is strongly influenced and pervaded by vested class interests. Education, provided by the state, promulgates the values needed to perpetuate the status quo (viz. the 'Enterprise Initiatives' in British higher education). Even the welfare state, to a true revolutionary socialist, is *counter-*revolutionary, because it buffers people from the extreme effects of capitalism and so dulls their senses from realising what the system is really like and is really doing to them: it promotes, in other words, 'false consciousness'. The state also enshrines hierarchy and privilege, sanctifying it in the monarchy, for example. It diffuses discontent, and makes people think that a class-stratified society is a 'natural' one, or encourages them to think falsely that they live in a classless society.

Habermas sees a problem for capitalism in all this, for he believes that rationality has grown in influence, generally, through history. The trend is for people to become converted to values of technological rationality, against which it becomes ever more difficult, he thinks, for false consciousness to be maintained. For in late capitalism so many of the illogicalities and contradictions which underlie the system and its supporting liberal ideology have come to the fore. It functions more obviously and repressively, as things get worse, as a class system of increasing domination and control – but seeks to represent itself, increasingly unbelievably, as a society of freedom, democracy, equal opportunity and progress for all (Kearney 1986, 231–5). Hence capitalism faces a 'legitimation crisis'. It depends upon an increasingly difficult maintenance of false consciousness amongst people.

False consciousness is at the heart of alienation. It involves you in believing as objectively true, natural or inevitable, a set of premises that run counter to your own interests, which are not really true, natural or inevitable. And it imagines that what is part or yourself and your society – self-created – is somehow not so: is external and uncontrollable and controls *you*.

This happens in capitalist production, where increased specialisation and division of labour involve increased manual and intellectual cooperation between labourers. But these powers and processes of cooperation, and their fruits, are alienated because they 'confront the labourer as the property of another' (Smith 1984, 51). They no longer appear to the labourers who produced them to be *their* products, under *their control*. Products, and the labour they embody, have become objects apparently existing outside their producers. Imagining this to be so, labour cannot use its own powers of cooperative production to the full, to create a society for its own maximum benefit.

Labour thus appropriated for capital accumulation by others is further alienated by the economic 'laws' to which it imagines it must conform. Though such laws are a product of a specific mode of production whose continued existence depends utterly on labour's willingness to be appropriated, they appear to the labourers as external forces which control *them*. So labour's products are turned against labour when they are converted to capital for someone else's use. That someone else can withdraw 'their' (really the labourers') capital, destroying industry and jobs. Or labour's efforts can be turned into machines which then replace labour, or technology which labour cannot understand, control or own (TV, electricity generation, weapons), or into economic relations which are given the status of an external god-created thing – i.e. are mystified and sanctified. Ask most people what are the 'laws' of supply, demand and price, or what is the 'invisible hand', and they will probably not be able to explain. But they will, almost certainly, believe that such imaginary forces must be obeyed and that there is no alternative to doing so. So Thatcher's cunning phrase about 'laws of economic gravity' implies that existing economic and therefore social arrangements have the same immutable and universal status as physical laws.

True consciousness would recognise that in fact they do *not* have such status: they are inevitable and universal only if the *premisses* of capitalism (e.g. about the main purpose of economic activity being to generate capital) are also accepted. But such premisses are ideologically produced, and different ideologies with different social arrangements will have different premisses on which to base their economic 'laws'. False consciousness at present is really bourgeois consciousness, because it supports the interests of the bourgeoisie. It is complemented by scientific ideology as a secular religion, because science's main job is to develop productive forces (Smith 1984).

Nowhere is false consciousness more developed than in *commodity fetishism*, where the products of human labour are turned into commodities: seemingly impersonal things which do not derive their importance from craftsmanship or social usefulness, but primarily from exchange value. They are reified. Their real value (that of the labour they embody) is distanced (alienated) and fetishised (i.e. made quasi-mystical) as 'exchange value'. Similarly the relationships of production through which they came about are hidden. Their existence or not seems to be a function of the operation of straightforward economic 'laws' of supply meeting demand – whereas in reality much 'demand' was really *created* by the bourgeoisie through advertising and marketing, as 'new wants, and new needs' – for the primary purpose of capital accumulation through appropriation of the labour of others. True consciousness would recognise these hidden meanings in commodities, and would reinstate commodities as an organic part of the labour that produced them.

To summarise: social relations take the shape of 'things', as in commodities (reification); these things are invested with powers of their own, apparently over and above the powers of those who produced them (fetishisation); they then react on and against the interests of those who produced them and (unconsciously) allowed them to be invested with this power, which has now seemingly become 'external', 'objective' and 'real', like, for example, 'economic laws' (this is alienation).

Marxism's materialist approach to alienation should not be misconstrued as suggesting that alienation has only material manifestations. For it should be recognised that there are non-material aspects of it all. There are the losses of creativity and communality already referred to, and there is submergence of the individual and the personality in the mass production and consumption culture. All these are deep concerns of Marxism, particularly neo-Marxism.

Erich Fromm developed them into a thesis which married the idealism of bourgeois psychology with Marx's materialism. He identified alienation as a 'fear of freedom' (Coleman 1984), or, put another way, a 'poverty of desire'. So that for many people 'freedom' means freedom to do, or want to do, very little – least of all to revolutionise society and put people in control of their own social and economic existence. This fear also extends to reluctance to express one's feelings, emotions and individuality in full.

People learn to repress these things, and therefore a side of their personality, through different modes of submission. There is religious mystical submission, whereby individuals lose their real selves in a god. There are authoritarian or sado-masochistic submissive relationships, where you give yourself entirely to another. And there is willingness to have capitalism tell you what your needs are.

It should be stressed that this neo-Marxian concern for the non-material side of people does not compromise Marxist materialism. It does not take the Freudian stance that sees capitalism and the anti-social impulses it fosters as a result of repression of the personality. Rather, it holds the reverse: these repressions and impulses ultimately stem from the specific mode of production of capitalism.

3.4 IMPLICATIONS FOR ENVIRONMENT AND NATURE

Root causes

Greens frequently declare that environmental damage is the result of 'industrialisation' (Porritt 1984), allied to wrong attitudes and values: especially those inherent in classical science, and perhaps also in Christianity and patriarchy (White 1967, Capra 1982). Add to this a dose of guilt, invoking 'greed', 'hubris' and original sin (Pepper 1991, 116–18) and there are all the ingredients of a powerful cocktail of false consciousness which makes humans in general and the *self* in particular the 'seventh enemy' (Higgins 1980). But as Parsons (1977) observes, 'We have met the enemy and it is us' is really a self-accusing and self-moralising abstraction amounting to mumbo-jumbo.

By contrast, a historical materialist, socio-economic analysis of capitalism demonstrates that it is not just individual 'greedy' monopolists or consumers who are to blame, but the mode of production itself: the pyramid of productive forces surmounted by productive relations which constitute capitalism. Parsons considers that within the system the share of 'blame' can be distributed according to the distribution of power, but this analysis runs the risk of forgetting that those who control and manage the military–industrial–government complex are as alienated as the most menial labourers in it.

It is more accurate to say that it is 'the way in which human "interference" with nature is managed under capitalism that is the cause of much land degradation and the appalling human consequences that stem from this' (Johnston 1989, 95). And poverty, an ultimate cause of much environmental degradation (UN 1987), is a necessary feature of capitalism 'to goad its people into a continued competitive striving' (Seabrook 1985, 37).

Hence Marxism emphasises the dynamics at work in the material production processes which cause environmental degradation. It also enables us to see how attitudes to nature are shaped specifically in capitalist development, in such a way as to facilitate exploitation. First, there is large-scale physical removal of people from the land in the development phase of industrial capitalism and associated urbanisation and factory-based production. This stems from a need for a surplus population to work in industry. It is combined with a requirement for amounts of capital to invest in new industrial production. Thus there is pressure to greatly increase agricultural productivity, through technological improvements – hence agricultural 'green' revolutions.

Despite appearances, neither in early capitalist Britain nor in the contemporary third world did or does capitalist agriculture expand in response to any perceived Malthusian pressures of population expansion (Johnston 1989, 84). Rather, its purpose is to get a return on investment: to accumulate wealth in commodity production. Furthermore, the objectification of land and its products, through commodification, makes a second, material, contribution towards a developing exploitative attitude: that of distancing of – objectification of – nature.

The ecological contradiction

The dynamics explored in Chapter 3.3 all combine to make capitalism *inherently* 'environmentally unfriendly', although how much this shows at a given time will fluctuate: profitable operations can afford greater environmental consciousness than can unprofitable ones.

There is the growth dynamic, which provokes continuing expansion of the resource base to meet the ever-extending range of goods and services on offer. This extending range is necessary because if consumption were not to rise, then, under competition, firms would have only the hope of taking market shares away from other firms in order to derive profits. As well, shares of existing markets would anyway be likely to fall as new competitors entered the market.

In fact *increasing* rather than steady profits are needed in order to increase capital accumulation, to reinvest in the hope of creating yet more capital. *By definition* this is what the system is about. Furthermore, because of the contradictions of overproduction and the falling rate of profit, producers must work doubly hard – to counteract falling demand, and to expand it instead – by creating new needs. So there is constant research, development, product innovation, advertising and marketing campaigns. This leads to a constant revolutionising of the instruments of production, as the *Communist Manifesto* put it – a process of 'creative destruction, which is embedded within the circulation of capital itself', creating constant innovation, which exacerbates instability, insecurity and in the end becomes the prime force pushing capitalism into periodic paroxysms of crisis (Harvey 1990, 106).

This is necessarily combined with proselytising at the superstructural level a very un-green philosophy, which, during the postmodern era, says Harvey, has mobilised fashion in mass rather than exclusive markets, emphasised the consumption of ephemeral services and saturated mass consciousness with manufactured images, notions of instaneity, temporariness and disposability (of commodities, values and stable relationships with things, people, places and nature). Schnaiberg (1980, 228–9) calls this cycle of rising production to meet increased consumption and increasing consumption to meet rising production the 'treadmill of production'.

Hence, what Quaini (1982) calls the 'ecological contradiction of capitalism' is produced, whereby the system continuously gnaws away at the resource base which sustains it.

> Capitalist production . . . develops technology and the combining together of various processes into a social whole, only by sapping the original sources of all wealth – the soil and the labourer.
>
> (Marx, *Capital*, 475, cited by Johnston 1989, 71)

At the same time, resource conservation, recycling and pollution control are discouraged in the free market by the drive to increase productivity and maximise surplus value. Obviously, such practices involve more costs, and it is good practice for firms to internalise returns but *externalise* costs – that is, to let society as a whole pay them:

[environmental] rehabilitation costs money, and that money can only come – in the absence of collective action [taxation and planning] – from what would otherwise be profits. In a competitive situation, the organisation that threatens its productivity by returning some of the income into non-profit-making activities such as restoring the landscape threatens its own viability . . . for a variety of altruistic reasons people will use part of their profit to enhance the landscape and so promote values other than those of the capitalist dynamo. But that can only be done if profits continue to be forthcoming. The need to make a profit from the exploitation of stock resources, in a competitive situation, means that there are strong pressures to minimise expenditure on clearing up the mess created by mining the resource.

(ibid., 101)

Indeed, 'mining' resources – stripping their assets without worrying about effects on future productivity – is an inexorable trend in capitalist economics, which externalise costs partly by discounting them to the future (Pearce *et al.* 1989) – the next generation is left to pick up the tab for present damage. It spawns what Johnston calls 'ecological imperialism', which prefers to exploit new land and resources because they offer great potential for initial profits and rapidly increasing productivity. This equivalent to slash and burn agriculture, but carelessly for capital accumulation rather than carefully for subsistence, began in medieval Europe with forest destruction and wetland drainage. From the sixteenth century it acquired global dimensions, culminating in dustbowls, desertification, rainforest destruction, and the imposition, in white settlement and colonisation of third world countries, of a land-use pattern geared to foreign markets. Marx believed that capitalist agriculture (and by implication other resource use) is not rational. It mindlessly leaves 'deserts' and degraded environments behind in the search for quick gains.

all progress in capitalistic agriculture is a progress in the art, not only of robbing the labourer, but of robbing the soil; all progress in increasing the fertility of the soil for a given time, is a progress towards ruining the lasting sources of that fertility.

(Marx 1959, 505–7)

Similarly, externalisation of costs can be seen in atmospheric, water and land pollution, in preferring road to rail transport, in throwaway products and packaging, and indeed in the 'rationalisation' of production via machinery – the social costs of resultant unemployment being charged to society as a whole. Both human and natural resources are being treated as a commons (Hardin 1968): but because there is no *sense* of communal ownership there is no sense that the costs as well as the benefits of their exploitation will be communally shared.

Countless examples arise every year of private firms externalising social and environmental costs, openly or covertly. Rarely, however, do they openly *admit* to doing so, as did National Power, Britain's newly privatised electricity generating company, in 1991. It reportedly decided to close the country's leading laboratory for acid rain research, at Leatherhead, despite earlier government pledges that electricity privatisation would not harm environmental work. A Labour Party energy spokesman commented: 'It's not in National Power's commercial interest to find out the environmental impact of their industry'. And, as if to confirm this, a National Power spokesman said that its cost-cutting review '. . . takes into account our changed status as a private business. *We can no longer justify things just because they are in the general national interest – that is not how a commercial company is run*' (*Independent on Sunday*, 14 July 1991, emphasis added). Such candour is both sickening and refreshingly honest, as is that from someone writing from the Glasgow Business School, and commenting on how businessmen had lied to the Department of Trade and Industry to get approval to take over a famous department store:

> Competitive commerce is not a game of cricket. Entrepreneurs owe a duty to their families, their backers and to themselves to maximise competitive advantage by all possible means, short of breaking the law. An entrepreneur who baulks at misleading some jack-in-office, with no legal come-back, is clearly derelict in his duty. Such a person is not fit to have control of the property of others, for he is liable to place personal whim or scruple before the duty of care towards the assets in his charge.
>
> (*The Times*, 17 March 1990)

More normally business people justify what is in their narrow sectional interests by claims that they are *in*, not apart from, 'the *national* interest'. Conversely, we find national politicians defending 'national' interests which are really those of big business, particularly when the latter has fed in money to their political campaigns. This was clearly illustrated when Western, particularly American, administrations blocked, resisted and stonewalled the following proposed measures in preliminary negotiations before the 1992 UN Conference on Environment and Development (the Global Summit): reference in any treaty to safety standards in developing new biotechnology products; endorsement of the Bamako Convention banning toxic waste shipments across international borders; agreement on deadlines to reduce CO_2 emissions; reference to the West's overconsumption as a cause of environmental degradation; a proposal that transnational corporations should accept environmental liability for the 70 per cent of world trade which they control (Vidal and Chaterjee 1992). Clearly, ahistorical claims to be pursuing the universal good are, like the 'invisible hand' concept, really highly ideological. There is seldom one 'national interest', just as there is seldom a 'common' sense. However, as with all preceding ruling classes, the bourgeoisie has always had the ability to generalise its own particular class view of the world as being a world view, a universal truth.

(Burgess 1978)

Such universalisation was always an important weapon in capital's search for legitimation and it is no less so over environmental issues. Hence, in post-war Britain the pollution which would be caused by a proposed private brickworks in Bedfordshire provoked little opposition or public scrutiny at the time: the Minister accepted the proposals on grounds of overweening 'national interest' (Blowers 1984).

Yet, as O'Connor (1991a, 13–15) points out, if capital *as a whole* were to behave rationally it *would* work for a more, though by no means complete, universal good, i.e. according to the long-term sustainability of the whole system, including the 'conditions of production in general and environment or nature in particular'. It would, in the form of the capitalist state, look out for

the interests of capital as a whole, and . . . follow policies designed to keep profits and investments and markets expanding, i.e. to keep costs of conditions of production low and the average rate of profit high. Such a state would have a rational education and health policy; a rational and sustainable environmental policy; and a rational urban policy. But we know that the state has none of these things.

Hence the cost of capital goes up. This is for

a *systemic* reason, namely, that *individual* capitalists have little or no *incentive* to use production conditions in a sustainable way – i.e. in the race to cut costs capital has impaired its own profitability [emphases added].

In thus demonstrating how there is no hope for a sustainable capitalism, O'Connor emphasises how the contradiction whereby *individual firms* behave ecologically (and socially) in a way which is against their own interests, long term and collective, (i.e. the tragedy of the commons) is inherent to the system. He also shows, therefore, why Hardin's conclusion that the solution to the tragedy lies in more privatisation of the commons is fundamentally unsound.

The ecological contradictions of capitalism make sustainable, or 'green' capitalism an impossible dream, therefore a confidence trick. As O'Connor again says (p. 15):

all the green consumption in the world will not change the fact that aggregate consumption must stand in a certain relation to investment for capitalism to work, and that aggregate consumption is not regulated by consumers but by the rate of profit and accumulation – and the limits of the credit system.

It must further be remembered that part of capital's general dynamic is to set goals and ideals for everyone which, in reality, are not attainable by most without destroying the system. Just as Seabrook (1989) demonstrated this in relation to material riches, so Hardin demonstrated it in relation to environmental quality. If people flock to a wilderness to be 'alone' with nature, they will destroy that solitariness which they sought. (Anyone who doubts this need only to try to experience spiritual bonding to nature through climbing Ben Nevis in Scotland during the August holiday season.)

So 'ideal' capitalism, which rewards all those who are part of it, may exist for one group, in one place, but only at the expense of far from ideal capitalism for others elsewhere. Western European and American levels of affluence (in general) are achieved on the back of a 'billion people living in absolute poverty' and all the other obscene inequalities shown in Table 1.2. But Marxist and ecological analyses both demonstrate that the more that populations from the second and third worlds join in this race to mythical riches by adopting capitalism, the more certain it is that *general* levels of long-term economic prosperity and environmental quality will drop. The principle of the prisoner's dilemma operates. Everyone can gain from the situation only as long as everyone is content to make small, incremental gains, and can trust each other not to go all out for maximum individual gain.

By this token the wonderfully ecotopian city of Davis, California, is founded on

> the process of reproducing on an expanded scale inequalities and privilege in Davis *vis-à-vis* the rest of the world, i.e. the process of keeping the city socially 'clean', which is WASP shorthand for keeping out factories, working-class people, blacks and other oppressed minorities.
>
> (O'Connor 1991a, 15)

And since environmental quality is linked to material poverty or affluence, increasingly as Western capitalism sustains and 'improves' itself to become the world's envy by siphoning off third world wealth, so its new-found 'greenness' will be achieved by making less privileged areas toxic waste dumps denuded of trees and soil. Peter Greenaway's metaphor in 'The Cook, the Thief, the Wife and Her Lover' is increasingly apt here. The elegant façade of a gracious cordon bleu restaurant is made possible only by a series of increasingly filthy and nauseating back rooms and kitchens.

Orthodox Marxists are sometimes thought to interpret all this as meaning that capitalism alone is responsible for environmental destruction, and we do read Marxist statements such as: 'There are real limits to growth today. But these are the products of capitalist society, not laws of nature' (Richards 1989). Historically the issue is more complex, however, and the fairly orthodox SPGB (1990) realises this, pointing out that a class of overseers of agricultural surpluses, with powers that were usable to limit some people's access to the means of production, appeared as soon as sedentary agriculture made production for surplus possible, and not specifically in capitalism.

And Grundmann (1991), a neo-Marxist who thinks (p. 51) that it is 'plainly wrong' to believe that capitalism is the main cause of ecological problems, considers that even if class society were abolished modern society would still suffer from alienation. This would mean that social and nature relations would still be seen as outside individual and community control (see Chapter 3.6). This, which is a *lack of human self-development*, is what, he says, ultimately causes ecological crises. But this merely underscores the most important fact for him, which is (p. 79) that even though many other modes of production apart from capitalism generate ecological disruption, only in an unalienated, *communist*, society will there not be bad environmental problems. Indeed, true communism *by definition*, he says, cannot have them. Grundmann cites evidence to suggest that this perspective in fact exists in neglected parts of Marx's theories. For instance, in *Capital* Marx blames soil exhaustion on ahistorical factors such as 'greed' as well as the historical capitalist mode of production. Again, Marx hovered between historicism and ahistoricism in discussing technology and division of labour, which may, inherently, de-skill and cripple people and harm the environment (ahistoricism, in *Manuscripts* of 1861–3) or do this only under capitalism (historically specific, in *Capital*). Grundmann eventually (pp. 200–1) goes for an interpretation which suggests that Marx invoked inherent, ahistorical characteristics in human society until it succeeded in reaching the stage of communism, whereupon it would be able to *control* its form and its interaction with nature. So historically technology has determined society and created social good and evils, but the reverse should and would be the case under 'social production' (communism), where technology will do no harm.

Capitalists and their liberal apologists excuse the economic and environmental disparities of Table 1.2 by increasingly vigorous recourse to the demonstrably false 'trickle down' theory of wealth (which says that if you allow an elite at the top to become as wealthy as it can, society will become richer and the good effects will diffuse downwards). And in true Malthusian fashion they put the blame for third world poverty on to the shoulders of the third world poor themselves, through the 'overpopulation' myth.

3.5 MARXISM AND THE POPULATION—RESOURCES ISSUE

The orthodox theory

The theoretical perspectives described above demand that current perceived problems of population numbers in relation to resource availability must be seen in a *historical* perspective; that is, in relation to specific modes of production. Marxists say that only animals and plants are bound by an 'abstract', that is an objective, unchanging, non-historical, law of population.

Where human societies have little power to change their environment population size *is* an important limiting factor on material wellbeing because of physical limits on resources. However, where societies can change and manage their environments the population size which can be sustained is determined more by social relations. Thus we can talk not of one universal law of population, but of a law of population which is specific to a given mode of production, like capitalism, and is bound up with social relations of production.

Marxists ask what a term like 'overpopulation' actually means: overpopulation in relation to *what*? And how do we know when there *is* overpopulation? We may say that overpopulation is evidenced by the existence of groups who do not have enough to eat (they are presumably the surplus people), which is what resource scarcity implies (Perelman 1979). But it does not really follow that this starvation is produced by 'natural shortages', i.e. an *absolute* inability of the earth to produce more food. Rather, the 'surplus' population may not be able to buy food simply as a result of the inability (or unwillingness) of an economic system to create enough jobs, and therefore incomes, or to pay enough to those who do work.

In fact, as Chapter 3.3 showed, it is fundamental to the nature of capitalism that wages must be kept as low as possible. To maximise capital accumulation, that is surplus value, wages must be kept depressed in relation to the exchange value of products, even when the latter may be increasing. One way to do this is to have an unemployed pool of people who will be ready to step in and do poorly-paid jobs if those already in them threaten to strike for more wages. Competition for jobs keeps wages low. This is not some nineteenth-century Dickensian scenario. Repeatedly in the 1980s and 1990s in Britain, unemployment has been used nakedly as a weapon to depress wages, and to argue against a minimum wage, and against strong unions.

The production of the pool, the *'industrial reserve army'*, is an inevitable outcome of the constant tendency for machines to be substituted for labour, *whatever is the absolute size of the population*. So the reserve army of the unemployed, produced by mechanisation in order to maximise capital accumulation, permits the expansion of the surplus value produced by labour, and also furnishes a pool of labour to draw on in times of boom. At other times, however, the members of the 'army' *appear* to be surplus to requirements, even though they fulfil this vital function for capital.

They cannot buy enough, so *for them* resources are scarce. And by keeping the wages of others depressed, this relative surplus population serves to give an ever-wider appearance of Malthusian 'starving, overbreeding masses'. But poverty is not said by Malthusians to result from this politico-economic process. Instead it is put down to ecological population 'laws', which are thought unavoidable except through the efforts of the 'overbreeders' to restrain their own behaviour. However,

The production of a relative surplus population and the industrial reserve army are seen in Marx's work as historically specific – internal to the capitalist mode of production. On the basis of his analysis we can predict the occurrence of poverty no matter what is the rate of population change.

(Harvey 1974, 231)

Harvey goes on to illustrate the value of a historical perspective on this problem. He examines the typically neo-Malthusian sentence

Overpopulation arises because of a scarcity of resources available for meeting the subsistence needs of the mass of the population.

He focuses on the words 'subsistence', 'resources' and 'scarcity'. The first, he says, has a relative meaning which is socially and culturally determined. Needs can be created, and are not purely biological, thus what constitutes 'subsistence needs' is internal to a particular society, with its particular mode of production. Subsistence needs therefore change over time. Similarly, 'resources' are things in nature which can be useful. Thus they can be defined only with respect to a society's stage of technical development – i.e. its *ability to use* nature's materials. As Spoehr (1967) says: 'It is doubtful that many other societies, most of which are less involved with technological development, think about natural resources in the same way as we do'. And 'scarcity' is not inherent in nature as neoclassical economists maintain; its definition is 'inextricably social and cultural in origin' because it can be assessed only in respect of what a society wants to attain in the first place (e.g. mere biological subsistence or high economic living standards). The original sentence can be therefore rewritten as

There are too many people in the world because the particular ends we have in view (together with the form of social organisation we have) and the materials available in nature that we have the will and the way to use, are not sufficient to provide us with those things to which we are accustomed.

(Harvey 1974, 236; emphases added)

This redraft makes population–resource problems no longer insoluble by any human means other than by altering our population numbers. For we could change the ends we have in mind, and social organisation (to deal with 'scarcity' where it exists), or we could change our technical and cultural appraisals of nature ('resources'), or we could change our views concerning the things to which we are accustomed ('subsistence'). For Harvey, the problem about focusing on how to change population numbers lies in the ideological implications of such discussions: 'Whenever a theory of overpopulation seizes hold in a society dominated by an elite, then the non-elite invariably experience some form of political, economic and social repression'. Who are the surplus people? Clearly it is not us, so it must be *them* – immigrant workers, racial minorities, the third world's starving masses and so on.

So one facet of the Marxist approach to questions of resource availability and subsistence is to look hard at these terms themselves and set them in their economic and social context, rather than investing them with notions of universality. One can readily see how, in capitalism, the idea of a need (and hence resource availability to fill it) is highly contingent on social relations of production. 'Need' is not usually expressed in terms of what would be socially useful to all of the people, but in terms of the aggregate of individual demands, or 'wants', expressed mainly by those *with appropriate purchasing power.*

Ryle (1988, 68–70) considers that Marx is right to point out how one set of needs is replaced by another when modes of production change, but wrong to apparently celebrate capitalism's productivist ethic as driving 'labour beyond the limits of its natural paltriness' and creating 'the material elements for the development of rich individuality' (*Grundrisse*, 325). William Morris, he says, follows a more appropriate eco-socialist line when he distinguishes between 'real' needs that support life and the unnecessary 'sham' needs produced by capitalism.

But Ryle also discloses a vital aspect of Marx's nature–society dialectic (see below), which is that

> at any moment the forms taken by quite ordinary and basic needs will be newly developed: everyday social reproduction incorporates a large quantum of historically new needs.

Indeed, are not telecommunication and wheeled transport 'basic' needs today?

Interestingly, to follow strictly the latter idea, of the historicity of resources, brings Marxism close to the free-market liberal analysis developed by Simon and Kahn (1984). This is that resources are *not* finite. It may or may not be true that absolute amounts of copper oxide or petroleum are declining, but this is not relevant, since we do not directly consume either. What we do consume are telecommunication – and this can now be done by means other than copper wires – and automobile travel – where the 'water' powered, 'pollutionless' electric car could be a future alternative to the internal combustion engine car.

Neo-Marxists and natural limits

Perhaps it is potential left–right ideological collusions that steer some eco-socialists – like those described in Chapter 3.1 – away from such strict implications of Marxist analysis. Benton (1989), for instance, believes that strands within Marx's economic analysis led him to under-theorise or leave out limits on the use, i.e. transformation, of raw materials, the fact that they and technologies all ultimately come from nature, and how naturally-given geographical and geological conditions cannot just be subsumed as 'instruments of production'. Marx's analysis was ideologically opposed to objective

Malthusian limits (Chapter 3.7 explains why) it viewed capitalism progressively for its accelerated development of productive forces (on which communism would have to be built), it shared the nineteenth-century industrial ideology, and it overridingly emphasised human intentionality.

Consequently, Benton thinks, Marx overestimated the role of human behaviour and underestimated the significance of non-manipulable nature. He did not concede, for instance, how in agriculture human labour changes the environment but it is nature that actually grows the fruits – labour merely regulates and reproduces natural processes. Furthermore, Marx's economics drew greatly on Ricardo, who emphasised labour's role in creating value without dwelling on the importance of natural-resource scarcity. Ricardo himself, however, did recognise that such scarcity played a role in value creation.

However, says Benton, there is a discontinuity between Marx's economics and his historical materialism. For the latter *did* persistently convey a view of human society as dependent on nature-given material conditions. It also accepted that different geographical endowments determine different ways in which people transform nature, so that historical materialism could justifiably be described as 'human ecology'. Marxism must thus now be reconstructed to overcome this hiatus.

In advocating this, Benton, in common with many other neo-eco-Marxists, seems generally not to take the trouble to *prove* the orthodox theory wrong. They accept, *prima facie*, that the ecologists are right, i.e. that ultimate limits to growth imply immediate and significant, rather than distant, impediments to human development even under socialism.

Parsons (1990), for instance, uncritically embraces the limits to growth thesis: 'Socialism should be informed by rationality and science', he says, 'and the queen of science is ecology'. Ecology tells us, he asserts, that a policy of matching population to resources is needed for Africa, which 'is short of food, fuel and many other basic necessities *now*'.

Pre-colonial practice

This is so, but historical materialism stresses the importance of realising that it was *not* so before Western colonisation. Then, 'third world' populations were in ecological balance, and harmony with nature was a feature of their lifestyles (Redclift 1986):

> We might do well to reflect that 'poverty' as it is known today was almost unknown in pre-colonial Africa. Although the pre-colonial era was not a Golden Age . . . there was no 'overpopulation' in the sense of a rate of population increase greater than the rate of increase of food production.
>
> (Omo-Fadaka 1990, 180)

And malnutrition or unemployment were rare until people began to be brought within the colonial orbit.

Keleman (1986) strikingly confirms this assessment, by contrasting two descriptions of Tigré province, Ethiopia, from 1901 and 1985:

The environs of Adowa [Aduwa] are most fertile, and in the heights of its commercial prosperity the whole of the valleys and the lower slopes of the mountains were one vast grain field, and not only Adowa, but the surrounding villages carried a very large, contented and prosperous population. The neighbouring mountains are still well wooded. The numerous springs, brooks and small rivers give an ample support of good water for domestic and irrigation purposes, and the water meadows always produce an inexhaustible supply of good grass the whole year round.

(Wylde 1901)

Shortly before I left Ethiopia I flew over large tracts of the desiccated provinces of Tigré and Wollo. For hours the picture below was unchanging: plains which formerly were described as the breadbasket of the north were covered in rolling mist of what was once fertile top soil; eddies of spiralling dust rose in the whirlwinds hundreds of feet into the air; stony river beds at the bottom of gorges a thousand feet deep showed not a sigh of water or new vegetation; and the grazing land at the top of the plateaux which the dried-out rivers dissected were as bald and brown as old felt.

(Vallely 1985)

A Marxian analysis of this difference relates it to the eclipse by capitalism of the pre-colonial mode of production: 'primitive' communism (or communalism).

In the latter, Omari (1990) tells us, common ownership of the land prevailed. The limited possession rights of individuals and families over specific portions of land were subordinate to the ownership rights over all land of the tribe or clan. Every member, dead, living and unborn, had ownership rights, and while individuals might transfer possession rights, no money changed hands.

Young (1990) confirms that this system also prevailed among aboriginal people in Oceania. He suggests that for colonialists to apply the term 'wilderness' to portions of land was an ideological confidence trick: it implied that no-one owned it, when in fact *everyone* owned it.

The material fact of common ownership translated into common attitudes and practices towards the land in Africa and Oceania. It was part of the community, in the sense that Leopold (1949) wanted land to be in modern America. It was held in trust for the next generations, and animals and birds were worshipped as totems and protected by tribal law, as were trees:

In traditional African societies, religious taboos and restrictions took the place of afforestation campaigns which are now being waged by governments like that of Tanzania.

(Omari 1990, 170)

Such attitudes were changed with the advent of the money economy, Christianity, Islam and Western individualism.

Capitalist penetration

The internationalisation of capital was and is an inevitable response, Heilbroner (1980) reminds us, to capitalism's internal contradictions. It is not an irrational process since it follows the dictates of capital as self-expanding value, but it leads to the irrationality of uneven development. Neither does the process produce regard for the health and welfare of indigenous people.

Superficially it appears to be a mutually beneficial symbiosis of peasant culture and capitalist economics, but the third world peasants (a true proletariat) are the weaker partners in the face of a bourgeoisie consisting of Western economic interests and an elite third world minority. What happens to the peasantry is a form of 'structural violence' (Johnston 1989, 95) – that is, violence caused indirectly through the effect of specific political, economic or social policies or institutions.

Thus the peasantry's communal land is privatised and brought into production for a capitalist world market. As such, it is constantly pressured for increased productivity. Therefore many peasants are driven off it; either into mushrooming cities, or to marginal land. This they are forced to overcultivate, because of their numbers, leading to soil erosion and desertification. Peasants who remain to farm the best land, and therefore remain in the market system, are forced to compete with other third world producers. To counter the steadily declining prices which their produce gets on the open market because of such competition, they are pressured to make further productivity increases, through investing in green revolution techniques and equipment – produced by the West. Productivity also demands high-tech infrastructure projects, such as roads and dams. Consequently large debts are incurred with the West (particularly in the 1970s when Western banks were anxious to re-invest Arab money deposited with them). Hence, yet further productivity increases are sought, to pay them off.

That Western colonisation has *always* been driven by economic motives is strongly suggested by the 'distinguished' figures cited by Goldsmith (1990).

We must find new lands from which we can easily obtain raw materials and at the same time exploit the cheap slave labour that is available from the natives of the colonies. The colonies would also provide a dumping ground for the surplus goods produced in our factories.

(Cecil Rhodes, founder of Rhodesia)

103

The colonial question is, for countries like ours which are by the very character of their industry, tied to large exports, vital to the question of markets. . . . From this point of view . . . the foundation of a colony is the creation of a market.

(Jules Ferry, speech to the French House of Deputies, July 1885)

We have spoken already of the vital necessity of new markets for the old world. It is, therefore, to our very obvious advantage to teach the millions of Africa the wants of civilization, so that whilst supplying them, we may receive in return the products of their country and the labour of their hands.

(Lord Lugard, British Governor of Nigeria)

The most useful function which colonies perform . . . is to supply the mother country's trade with a ready-made market to get its industry going and maintain it, and to supply the inhabitants of the mother country – whether as industrialists, workers or consumers – with increased profits, wages or commodities.

(Paul Leroy-Beaulieu, *De la Colonisation chez les Peuples Modernes*, 1874)

Goldsmith further notes how nations like China were forced into trade with the West by gunboat diplomacy. And, since the 1944 Bretton Woods conference, the richest nations have redoubled efforts to bring (former) colonies into the world trade system, specifically to avoid another 1929-style crash, caused by overproduction at home. The International Monetary Fund, the World Bank and the General Agreement on Tariffs and Trade (GATT) were instituted for this purpose. They have encouraged and coerced third world countries to lower their import quotas and tariffs, which protected their fledgling industries, to devalue currencies, making their exports cheaper and Western imports dearer, to cut welfare expenditure and to capitalise agriculture via Western machinery and agro-chemicals.

Transnational corporations already command 80–90 per cent of the trade in tea, coffee, cocoa, cotton, forest products, tobacco, jute, copper, iron ore and bauxite, but they want more. Hence in 1990 they pressured the US government to insist, in the eighth (Uruguay) round of GATT negotiations, on further opening up and deregulating trade with the third world – such deregulation to include allowing only minimal pollution control measures.

Raghavan (1990) calls this an attempt to 'recolonise' the third world, noting that while American finance capital interests want third world trade barriers removed, its industrial capitalists want the West's barriers to remain, so protecting manufacturing industry against third world imports. This protection includes guarding Western rights over 'intellectual property' so that the third world cannot make cheap copies of Western products, thus dividing the world into knowledge rich and knowledge poor.

Ritchie (1990) shows how the US's GATT demands for third world agriculture have great environmental implications. Reducing domestic farm support programmes would mean less encouragement, direct and indirect, for farmers to protect the environment. Disallowing limitations on agricultural imports to the West would encourage forest exploitation for wood and beef. And making environmental and health safety standards uniform will encourage those standards to be minimal.

'The Uruguay round', says Khor Kok Peng (1990) 'can be seen as the transnational corporation empire's way of striking back at global demands for legislation to tackle ecological concerns and the third world's demands for global economic justice'.

By exposing the capitalisation of agriculture in Central America to class analysis, Faber (1988) shows how such ecological degradation is not merely inevitable: it is *necessary* to capitalism. The latifundia are the large capitalist estates and commercial farms, run by a bourgeoisie propped up by the military and international capital. They accumulate capital in a 'disarticulated' (disjointed) way: selling agricultural products abroad for relatively high returns to a rich market, but using low-cost home labour on their farms. This labour consists of peasants earning a paltry income on top of the basic food afforded them through their minifundia – subsistence farms on marginal land. In order that minifundio peasants will continue to be willing to provide this extra labour, at below subsistence wages, the subsistence sector must remain underdeveloped and not able to produce enough. This amounts to class war on peasants and workers, ensuring the 'ecological impoverishment of the people, which forces peasants to engage in wage labour [and this is] functional for the peasant model of development' (ibid., 42). Such ecological impoverishment readily stems from the land tenure arrangements, whereby the spread of cattle ranching dispossesses the subsistence classes, who are forced to intensify production on their marginal land.

But such processes can also lead to peasant-based agrarian reform and/or revolutionary movements, perhaps reflecting Marx's original hypothesis that continued oppression and enlargement of the proletariat would lead to the 'contradiction' of a developing revolutionary class consciousness. This would eventually be turned against the bourgeoisie whose actions had triggered it off, so the bourgeoisie would have become their 'own gravediggers'.

Gudynas (1990) also detects a class dichotomy within Latin American environmentalism. It has a managerial-technocentric group, the capitalist middle and upper classes, and 'antihegemonic' grass roots groups, of feminists, anarchists and liberation theologists who link human and nature rights, invoking the Franciscan tradition.

Gudynas's preferred solution, of sustainable 'ecodevelopment' involving bioregions, alternative technology and quality-of-life economics, would seek to reconcile these two classes. A more orthodox Marxist approach sees this as impossible. Thus Slapper (1983) insists on the need for a third world socialist revolution, ten years after which, he maintains, stable (i.e. ecologically sound) output would be possible. Research and technology would expand large-scale production of basic resources, with more diverse, local, small-scale production supplementing it. A corollary would be withdrawal from world capitalism: there would be no buying, selling or money and national boundaries would be destroyed.

While many greens might argue that this approach is based on an 'outdated' analysis, there are many aspects of what is happening now in the third world which echo the processes that Marx perceived in nineteenth-century Europe. Thus World Bank-funded 'development projects' in countries as far apart as northern India, Brazil (Rondonia) and Indonesia lead to certain common effects on local societies and economies (BBC 1987b).

These involve a change in the mode of production from 'primitive' communism to capitalism. Relations of production change from reciprocity to competition. Ownership of the means of production (especially land) becomes concentrated into fewer hands. The products of nature acquire cash value as opposed previously to use value (and sacredness). The benefits of the projects (roads, hydro-dams and transmigration) go to an elite minority in the West and third world, while the disbenefits accrue to a proletarian majority – they are, also, increasingly an urbanised proletariat. The projects themselves are based on short-term economic thinking: costs of environmental damage (silting rivers, destroyed rainforest) are discounted to future generations. Indigenous cultural variety is replaced by a hegemony of consumerist values and American trash culture.

Publicly the World Bank now regrets such malign environmental and social influences. But whether it has privately changed its outlook is rendered doubtful by a garbled internal memorandum written by its chief economist, Lawrence Summers, in December 1991, and leaked to the *Economist* (8 September 1992, 82):

> Just between you and me, shouldn't the World Bank be encouraging more migration of the dirty industries to the LDCs? I can think of three reasons:
>
> (1) The measurement of the costs of health-impairing pollution depends on the foregone earnings from increased morbidity and mortality. From this point of view a given amount of health-impairing pollution should be done in the country with the lowest cost, which will be the country with the lowest wages. I think the economic logic behind dumping a load of toxic waste in the lowest-wage country is impeccable and we should face up to that.
>
> (2) The costs of pollution are likely to be non-linear as the initial increments of pollution probably have very low cost. I've always thought that under-populated countries in Africa are vastly *under*-polluted; their air quality is probably vastly inefficiently low [*sic*] compared to Los

Angeles or Mexico City . . . [he means that their pollution levels are low].

(3) The demand for a clean environment for aesthetic and health reasons is likely to have very high income-elasticity. The concern over an agent that causes a one-in-a-million change in the odds of prostate cancer is obviously going to be much higher in a country where people survive to get prostate cancer than in a country where under-5 mortality is 200 per thousand. . . .

But so far the groups affected by World Bank projects have shown insufficient class consciousness and organisation to counter such crass, economistic imperialist thinking. Perhaps this is partly because they are spatially dispersed. In fact there is a *false* consciousness, whereby local people sell their labour for low wages, to build the projects that will destroy their communities. Whether a third world revolutionary consciousness might develop, as has not generally happened in the West, will be considered further below.

In turn, Western false consciousness about the third world is amply illustrated by Band Aid and other mass-media charity events. The cloying self-congratulatory tone of the 'generous' stars melds with the frenzied pointlessness of the supporters' sponsored activities to produce a rich cocktail of hypocrisy. Purged of some of their guilt, the participants then return, none the wiser, to lifestyles and politics (masquerading as non-politics) that create the very situation which their 'charity' had sought to alleviate. There are few exceptions, although the analysis of Bob Geldof himself, who started it all, seems to get closer with time to the Marxist perspective outlined above.

3.6 THE SOCIETY-NATURE DIALECTIC, AND ALIENATION FROM NATURE

The dialectic

Much of the present debate about the society–nature relationship tends to be couched in the language of classical science, even amongst those who want to reject that science and its values. This language talks of broadly deterministic relationships between separate entities – A causes or controls B, B controls C, etc. – even if those entities are linked in systems. Technocentrics may argue that humans do and should *control* nature: ecocentrics that natural limits do, or should, *constrain* human activity.

By contrast, Marxists offer a dialectical view of the society–nature relationship. This holds, first, that there is no separation between humans and nature. They are part of each other: contradictory opposites, which means that it is impossible to define one except in relation to the other (try it!). Indeed, they *are* each other: what humans do is natural, while nature is socially produced. Second, they constantly interpenetrate and interact, in a circular, mutually affecting relationship. Nature, and perceptions of it, affects and changes human society: the latter changes nature: nature, changed, affects society to further change it, and so on.

As Parsons (1977, 3) describes it, humans dialectically interpenetrate with the rest of nature via their perceptions, reflections and enjoyment of it, through *production*. Smith (1984, 16) says that the labour process is the 'motive force' of the metabolic society–nature interaction. In labour, humans incorporate their own essential forces into natural things, which thereby gain social quality as use values; so 'nature is humanised while men are naturalised'. The medium by which human labour interacts with nature is technology: therefore the characteristics of our technology also tell us something about the characteristics of our relationship to nature.

This process, of production, changes nature – the more so where advanced technology is used – and it also affects how we think about nature. Thus is nature *socially* produced. Uses and perceptions of nature vary with modes of production. Under capitalism, unlike previous modes of production, nature is changed to get exchange value as well as use value; and nature thereby tends to become objectified in the form of commodities.

Some Marxists draw attention to Marx's recognition of human dependence on nature as one of the forces of production, as evidence of his 'sensitivity to the objective existence of ecological laws' (see Chapter 3.3 and Parsons 1977, 19). Taking its cue from this the SPGB (1990, 1) accepts that there are laws of ecology, which are being broken, and 'humans are part of nature and cannot permanently defy its laws' (p. 23). Marx also recognised the priorness of an 'external' or 'first' nature, that gave birth to humankind. But humans then worked on this 'first' nature to produce a 'second' nature: the material creations of society plus its institutions, ideas and values. This process, as Bookchin (1987) stresses, is part of a process of *natural* evolution of society. Smith stresses how, under capitalism, as soon as any 'second' nature is produced, the human relation with it is mediated by exchange as well as use values. Thus, how we duse nature relates to the *expense* of various possible uses.

Whether there is any meaning left in the distinction between 'first' and 'second' nature in any place that humans can reach is a moot point (see below). Certainly Marx pointed out that:

> Animals and plants, which we are accustomed to consider as products of nature, are, in their present form, not only products of, say, last year's labour, but the results of a gradual transformation, continued through many generations, under man's superintendence and by means of his labour.
>
> (*Capital* I, 181)

> The nature that preceded human history . . . today no longer exists anywhere. . . .
>
> (*German Ideology*, 63)

In these days of concern about global atmospheric modification the truth of Marx's observations, cited by Smith, becomes clear on but a moment's reflection.

Marx's organicism and monism

Marx's green critics sometimes refer to him as Cartesian or 'mechanistic' (seeing, after the manner of classical science, the world as a machine, and reducible to parts that can be assessed objectively). However, his concept of the society–nature dialectic appears to be, in reality, deeply organic (seeing them both as making up one organic body) and monist (physical and mental phenomena can be analysed in terms of a common underlying reality). As Capra (1982, 216–18) suggests:

> Marx's view of the role of nature in the process of production was part of his organic perception of reality. . . . This organic, or systems view is often overlooked by Marx's critics, who claim that his theories are exclusively determinist and materialistic. In dealing with the reductionist economic arguments of his contemporaries, Marx fell into the trap of expressing his ideas in 'scientific' mathematical formulas that undermined his larger socio-political theory. But that larger theory consistently reflected a keen awareness of society and nature as an organic whole, as in this beautiful passage from the *Economic and Philosophical Manuscripts*: 'Nature is man's inorganic body – nature, that is, in so far as it is not itself the human body. "Man lives on nature" means that nature is his body, with which he must remain in continuous intercourse if he is not to die. That man's physical and spiritual life is linked to nature means simply that nature is linked to itself, for man is part of nature . . .'.

In fact the dialectic is more than just a 'systems view', but it has been described by Parsons (1977, 5–7) in terms of which Capra would much approve. It asserts, he says, the ultimacy of space–time events interacting with other space–time events, to unite and conflict with them, to change and be changed by them, to develop, decline and pass away. It is a vibratory interactive process applying to all kinds and levels of material existence – all physical and biological orders:

> From this perspective plant and animal organisms are in mutual, continuous and transformative relations with their immediate environments in a universe of natural systems of things and events in dynamic equilibrium . . . ecology is the application of dialectics to living systems and dialectics is the generalisation of the method of ecology from living systems to all systems.

Understanding, in dialectics, comes through the concept of 'fields', says

109

Parsons, which are systems of order consisting of unstable entities in definite positional relationships to other unstable entities. They are

regions of energetic activity differentiated into particular 'individuals'. . . . Every individual is in turn a field of subindividuals and at present writing no ultimate individual or partless particle is known to science. Individuals are in some sense discontinuous but are related to other individuals by the continuity of physical inheritance from the past and by lines of force in their fields binding them together in space–time relations.

This part of Parson's book is, unusually, not well supported by quotations from Marx and Engels themselves.

However, Heilbroner confirms the monistic nature of Marx's dialectical approach to knowledge, and, again, parallels with the physicists' approach are notable. Dialectics hold that the *act* of enquiring shapes as well as discovers knowledge, hence to study 'reality' means studying not what 'is', but what we make of it (praxis). Dialectics hold that all reality is ultimately motion, not rest, so to depict things as static and unchanging disregards their essence. Furthermore, all change involves not mere movement but an alteration in quality, and there is a universal changefulness in things, which Hegel called 'contradiction'. This means that the essence of reality consists of conflicting processes (thesis and antithesis) – which coexist disruptively and eventually resolve themselves in a new synthesis. This is how successive stages in history unfold (see Chapter 3.2). The society–nature relationship in this schema involves, first, an original undifferentiated unity between the two in early modes of production. Then they are fragmented into antagonistic forces under capitalism. Then, in communism is a new synthesis is reached: a new unity in which the two are differentiated but not antagonistic.

Historical development – changes in the mode of production – also involves a dialectic between town and country:

The antagonism between town and country begins with the transition from barbarism to civilisation and runs through the whole history of civilisation to the present day.

(Marx and Engels, *German Ideology*, vol. 1)

They thus saw town and country each as the repository of different material circumstances and values – countryside generally representing the status quo and the town being where new forces and groups who were not well-served by the status quo developed. The division of labour was essentially developed under industrialism, fed by the siphoning-off of people from the countryside. The economic, political and social consequences of concentrating productive forces into cities have been massively significant for capitalism.

Part of the resentment against, and escape from the worst aspects of, these consequences, has centred on the countryside, in the romantic movement, and in modern suburbanisation and commuter movements. Another part has centred on town and factory, in the form of strong, organised trades unionism. Marx and Engels also saw the town–country 'antagonisms' as the foundation of class distinctions (surplus agricultural production being taken to towns and their management being overseen by a separate class from those who produced them). And they held that the existence of the town implied a need for 'administration, police, taxes, etc., in short of politics in general'. Under capitalism, the city has pre-eminently to be seen in terms of its function of facilitating production, exchange and consumption. All of these functions and roles of town and country seem to be highlighted again in the development of capitalism in the third world. Here, attitudes to nature and its conservation may differ significantly between the two.

Lastly, the contradictions which are at the heart of the dialectical process cannot be understood simply through logic. Logic would hold that the contradiction of human is non-human, and humans are self-contained while what is non-human is also self-contained. Dialectical thinking, however, is essentially relational, and would say that the opposite of human is nature, and without the idea of one we cannot form the idea of the other. Certain processes are 'contradictory', not because they are chance opposites but because they unfold in ways that are *both integral to and destructive of* the processes themselves (viz. the expansionist drive is both inherent in capitalism, and will destroy it, through the ecological contradiction).

Thus, dialectical relationships are not apprehended merely by linear rationality, but also by intuitive insight. This often results in a language of paradoxes which 'defy the syntaxes of common sense and logic' (Heilbroner 1980, 58). For Capra, the fact that physics does the same makes its findings compatible with mysticism. Marxists, however, while they do not eschew intuition, subjectivity and the non-material, do reject 'understanding' through mysticism on grounds (which Capra would deny) that it alienates (separates) the subject who is doing the thinking from what is being contemplated. We will take this up again below.

Mutual transformation

An important aspect of the society–nature dialectic says that when humans change nature, through production, they also change *human* nature, i.e. themselves:

> Man sets in motion arms and legs, head and hands – the natural forces of his body – in order to appropriate nature's productions in a form adapted to his own wants. By thus acting on the external world and changing it, he at the same time changes his own nature. He develops his slumbering powers and compels them to act in obedience to his sway.
>
> (Marx, *Capital*)

111

Thus, through learning how to farm nature's products, we changed ourselves from nomadic hunter/gatherers to sedentary people. Through learning how to manufacture things we changed ourselves to an industrial society. Agricultural and industrial societies, and the individuals in them, are *qualitatively* different from each other, and from what preceded them, as indeed may a 'post-industrial' society be from an industrial one. As our ability to use resources has grown we have developed new needs: housing, energy, telecommunication. As we have changed our power to do things (e.g. via transport, computers) we have changed the things we want and need to do (e.g. travel).

This interaction is not just material. Through changing nature and making things, we have changed ourselves into creatures who can appreciate the *beauty* of what we create; buildings, machines, art. We have developed our subjective senses – our feelings and emotions, for example. So: 'The dialectic is made clear; as the world is increasingly more humanised, so too are the senses humanly developed as a social process'. Included in all this are our *intellectual* senses. As we transform nature, we get to know nature's laws in order to transform nature more effectively and usefully. As this happens we develop our own intelligence.

So as history develops, humans and societies naturally evolve through the processes and results of production: their principal material way of reacting with nature. As power to transform nature develops, so needs develop. Thus *needs and the power and resources to meet them are historically produced*: they change through history and in different cultures. The 'need' for oil was not a need until we developed the technology to use it: we had no need for quartz until we developed technologies for making glass, silicon chips and so forth (see Chapter 3.5).

The evolutionary nature of this mutual transformation can, overstated, become crude historical determinism. Thus Laptev (1990, 124) describes it as a 'geological' process: an ascent from inferior to superior which is 'historically inevitable'. It is progress towards transforming the biosphere to the 'noosphere': a term borrowed from Teilhard de Chardin which originally meant 'an envelope of thought around the planet'; a new 'thinking layer which emerged at the end of the tertiary'. Under the Russian scientist Verdanski this idealistic concept was transformed to signify a material envelope of the earth, changing under human influence. This is how Laptev uses it, adding that it will represent a genuine society–nature dialectic, rather than a harmful conflict, only when capitalism gives way to socialism and then communism: 'the genuine resolution of the conflict between man and nature and man and man'.

Others portray this process of mutual transformation in a more refined way, to detail its non-material as well as material aspects. Winner (1986), for instance, describes how using tools and wood can develop the 'human qualities' found in the activities of carpentry, or how employing the instruments and

112

techniques of music making makes a musician a particular *sort* of person, not just a producer of notes.

And, as societies, we collectively employ different technologies to change nature, with clear political results. There are *inherently* authoritarian and repressive technologies, like nuclear power, that require an undemocratic centralised political system ('Seamen contracted to work on ships carrying waste for British Nuclear Fuels have been ordered to say who they have been living with for the past five years or face the sack. . . . The company said the checks were needed in the interests of national security . . .', *Guardian*, 18 October 1991). Conversely, solar power is most compatible with democratic decentralisation. To say this is not to be technologically determinist, since society is ultimately free to choose the technologies it wants.

All this echoes Lewis Mumford's (1934) eloquent and long exposition of how the society–nature dialectic develops to produce a technological society as a culmination of natural evolution, and how this produces new kinds of individual and society. Creating the machine, as a transformation and transformer of nature, he argues, has brought new moral and cultural values to the society that has assimilated it into its nature.

Thus an interest in the factual and practical is no longer considered with snobbish disdain. Taboos of class and caste are no longer so conclusive and definitive. Machines have, too, added a new aesthetic cannon to the arts, and have fostered techniques of cooperative thought and action. Capitalism originally liberated such cooperation: now it perverts and suppresses its further development.

Nature itself is transformed into a human work of art, whose beauty and wonder are augmented by the quantitative and analytic appreciation which a machine culture induces. The human imagination has also been enlarged by scientific, technological fantasising, and by a new machine aesthetic: of cranes, skyscrapers and microscopes: 'There is an aesthetic of units and series, as well as an aesthetic of the unique and unrepeatable' (Mumford 1934, 82). Cubism was the first style to reflect the beauty of the machine. Photography enhances our appreciation of pure form in nature, while films bring distant environments near, recreating in symbolic form a world that is beyond our perception.

Mumford goes on to extol the new environment: 'Man's extension of nature . . . the elements of this environment are hard and crisp and clear: the steel bridge, the concrete road . . .' (ibid., 105), and he admonishes the arts and crafts movement (Morris, Ruskin) for lacking 'the courage to use the machine as an instrument of creative purpose' (ibid., 97).

The technocentric triumphalism over nature in all this is mitigated through Mumford's dialectical outlook, which asserts that in the new, essentially communist, 'neotechnic' society which modern machines foster we will go beyond the old 'palaeotechnic' approach of simplifying the organic to make it intelligible, to complicating the mechanical to make it more organic and therefore more effective, because harmonious with our living environment.

Mumford's organic and anarchist–communist ideology precisely pre-echoes by about forty years, communalist ecocentrism: much of Schumacher's seminal work seems to have come from the pages of Mumford. And Mumford moderated his own technocentrism after World War II, claiming that the machine world

> has isolated its occupants from every form of reality except the machine process itself . . . all forms of organic partnership between the millions of species that add to the vitality and wealth of the earth are either suppressed entirely from the mind or homogenised into a uniform mixture which can be fed into a machine.
>
> (cited in Guha 1991)

Alienation from nature

The conception of nature in Marx, then, is not as a mere stock of economic goods (a technocentric view), nor as a source of intrinsic worth or good (a deep ecology view), nor as an endangered ecosystem (tragedy of the commons survivalism). It conceives of nature as a *social* category: though there was an 'objective' nature, it has now been reshaped and reinterpreted by one aspect of itself; human society.

Since alienation means separation from aspects of the self (see section 3 above), then alienation from *nature* means a failure to conceive of nature *as* a social creation. Vogel (1988, 376–7) clearly conveys the meaning and implication of this failure:

> In discussions of alienation from nature we fail to recognise that the environment we inhabit is an environment of objects built by humans. . . . There is not a single object in my environment that is not literally a human object.

'True wilderness' is very rare, and where it exists it is highly *artificial*: a piece of 'nature' withdrawn from that natural order in which human transformative activity plays such a crucial part. It follows that if some people inhabit an ugly or dangerous environment, this means that society's own acts (in creating its environment) have remained 'powers over and against us'. We are alienated from our own creation because we have not yet exerted conscious social control over the processes – *our* processes – which created the environment, in order to make it more desirable.

This becomes even clearer when Vogel contrasts Marxist and deep ecology concepts of 'alienation from nature'. For Marx, overcoming alienation from nature means asserting its human-ness, through abolishing its sham externality and controlling and planning its use for *all* society. For deep ecologists, overcoming alienation means asserting the naturalness of humans by 'living in harmony with environment', effectively by admitting the power which natural laws (e.g. carrying capacity) have over us, and by 'living lightly on the earth' through trying *not* to transform it. For nature is the source of worth (it 'knows best') and it will be endangered unless we follow its rules. This view, by seeing human activity as encroachment on or violation of nature

> seems to be curiously guilty of just the sort of dualism it ascribes to the project of dominating nature. Somehow the activity of humans in transforming their environment, *alone of all other species*, is 'unnatural'.
>
> (ibid., 379; original emphasis)

Vogel might well be thinking of ecocentric material like Greenpeace's 1990 recruiting leaflet, which accuses 'man' of multiplying 'his numbers to plague proportions . . . and now stands like a brutish infant, gloating over this meteoric rise to ascendency'. McKibben (1990), from his rural retreat threatened with development, echoes such sentiments, not only noting the 'death of [first] nature' but quite clearly bemoaning it. Deep ecologists would have us revere nature and preserve it, to acknowledge our 'one-ness' with it. But, as Bookchin (1987, 248–9) also points out, such worship actually *mystifies* nature, placing humanity *far apart* from it. It is a 'Supernature with its shamans, priests, priestesses and fanciful deities'. Reverent mystification really *separates* us from nature: we the 'impotent and terrified mortal before a jealous and angry god' (Vogel 1988, 379). Gaia is an inhuman force we cannot change but to which we must adjust for our survival. And this is just how Lovelock (1989, 212) presents Gaia: 'She is stern and tough, always keeping the world warm and comfortable for those who obey the rules, but ruthless in her destruction of those who transgress'.

So deep ecology's view of alienation from nature really rests on a *dualistic* conception of the human–nature relationship: a conception it is supposed to reject. Marxism's dialectic, however, is truly monistic, and so, as Vogel says: 'The question is not whether what we do "accords with nature", it is whether we like what we have wrought'. Or, as Smith puts it, the first question is not whether or to what extent nature is controlled: this uses the dichotomous language of first and second nature. The question is *how* we produce nature and *who* controls this production of nature.

Idealism and materialism

Parsons (1977) demonstrates how, in several aspects, this dialectical conception of the society–nature relationship avoids the extremes of idealism and materialism.

First, as we have noted, it rejects the mystifying and religious, passive idealisms in feudalism and capitalism which holds with the existence of a *supernatural* realm. It also rejects philosophies that make nature 'illusory': i.e. purely a product of human consciousness. And, adds Smith (1984, 37), it rejects the 'severe abstraction' of human–nature unity in modern physics. By contrast

'The unity of nature implied in Marx's work derives from the concrete activity of human beings and is produced in practice through labour', in other words, it is founded on materialism.

But, second, this does not imply a gross, vulgar materialism: that of capitalism, which, says Parsons, 'bulldozes and creates a denatured world', dead and neutral, and is founded on the exploitative and dominating ethic of classical science. This Marx sees as a real contempt for and a practical degradation of nature (Marx, *Early Writings*, p. 7, in Parsons 1977, 17).

And, third, while it does not deny the importance of external material objects and objective physical laws that we violate at our own peril, it also emphasises how interaction with our physical world develops our non-material personality (cf. Mumford).

It is an approach to the nature of being, therefore, that refutes both Hegelian idealism and the atomic materialism of classical science that serves capitalism. Parsons says:

> Man's transcendence does not carry him beyond the natural universe. Its multiple-loop feedback system of dialectical theory and practice does not violate the physical laws of nature: rather, it differentiates, integrates and refines them at a new level.
>
> (1977, 54)

> A correct view of it [nature] and man's place in it would do away with the traditional dualisms of man and nature, subject and object, fact and value and the like. It would exhibit the intimate interdependencies of nonhuman and human nature and their transformation.
>
> (ibid., 50)

Grundmann (1991) also takes this view of Marx's perspective on nature. Marx charts a 'third way' between the dualisms of dominion or stewardship, and instrumental versus intrinsic value. He does embrace both dominion and instrumental value, as ecocentric critics often allege. But the latter's interpretation of what Marx *means* by this is incorrect, says Grundmann. For, taking his position from Bacon's original project, Marx realises that nature is only harnessed by *obeying* its laws. 'Domination' does not, then, mean breaking an alien will, but, through cooperation, being able to *steer* nature (pp. 57–61). 'Nature in Marx is not anthropomorphous. Nature has no end in itself, it is man who imposes his ends on it. In order to do so, however, he has to respect the laws of nature'. As the SPGB (1990) puts it '. . . to talk about a struggle with nature is quite inappropriate. The utilisation of nature by humans to supply their needs involves cooperating with nature, not battling against it'. Conversely, says Grundmann (1991, 92), 'a society which does not take into account the repercussions of its transformations of nature can hardly be said to dominate nature at all'. Midas's type of power, for instance, was self-defeating. But in communism there will be 'domination', which means that all relations with nature will be under *conscious, common, human control*. Nature's value will be 'instrumental', but not in an economic utilitarian way, which 'neglects the fact that people might not want the extinction of an animal species, even if its actual financial value were negligible' (p. 67). Human 'use' will greatly involve moral, spiritual and aesthetic values – but *human* ones, not imagined 'intrinsic' ones emanating from an external, worshipped nature with its own mystical, unapproachable teleology.

First and second nature: use and exchange value

The idea of first and second nature was mentioned above. It pervades Western thought and is also part of Marx's dialectical view, according to Schmidt (1971). However, both Smith (1984) and Redclift (1987) are at odds with Schmidt's interpretation of Marx.

Schmidt is charged with incorrectness in ascribing two concepts of nature to the Marxian dialectic. One sees, in pre-bourgeois society, humans and nature truly unified; the first creating use values from the second 'naturally' and without the irreversible destruction of first nature. Humans and 'first nature' were one – not distinguishable as subject and object. But in bourgeois capitalism the dialectic of humans and nature consist of a consistent interaction of two relatively separate entities, as subject and object – humans dominating nature – to produce a second nature which cannot revert to first nature (the 'damage' is irreversible). Schmidt says that this second nature consists of commodities produced for exchange value, and that the exchange value of a commodity in fact has no natural content at all.

But Smith believes that there is only one concept of nature in the Marxian dialectic, and that is the monistic one (described previously). Nature, through productive processes and relations, is socially produced, and this was the case in pre-bourgeois society as much as it is in bourgeois society, which now turns nature into commodities:

> Exchange value falls within the realm of nature as soon as a second nature, through the production of commodities, is produced out of the first. The relation with nature is mediated by exchange value as well as use value determinants.
>
> (Smith 1984, 46)

And under capitalism virtually every use value now has a price tag – an exchange value. Everything is a commodity, even amenity and aesthetic enjoyment are pre-packaged and priced as commodities in tourism. So under capitalism it is no longer valid to distinguish between first and second nature – there is none of the first left. All of first nature is commoditised, so all of first has become second nature.

Smith's objection to Schmidt is bound up with the latter's adherence to the tenets of the neo-Marxist Frankfurt School of critical theory. This reacted against what it saw as vulgar economism in Marx, and therefore against arguments which ascribed 'destruction and domination' (i.e. production) of nature specifically to an *economic* mode of production – i.e. to production for exchange under capitalism. The implication of this would be, as discussed, that changing the mode of production, through revolution, to socialism/communism, might produce a different – more mature – facet of the dialectical relationship with nature, in which, as Mumford says, society is more organicised and nature less mechanised. The Frankfurt School did not believe this. In fact it eventually became anti-revolutionary, apolitical and ahistorical: asserting that domination and destruction of nature (as an object) would continue under socialism and was not ascribable to just one mode of production. Production, as 'industrialism', rather than *capitalist* production, was to blame for domination of nature and people, and the fragility of their existence. This position, which appeared in Horkenheimer's and Adorno's writings on which the Frankfurt School was based, does seem to reject Marxist historical materialism. Furthermore it reacts against an economic determinism and a technological triumphalism which it detects in Marxism but which Smith says is not inherently there, though it might be found in Engels, and Stalinist interpretations of him.

The Frankfurt School's critique does very much reappear in that of the contemporary mainstream and anarchist green movements. And as we examine below, the Frankfurt School has further common ground with the greens, in replacing the centrality of class struggle with a 'bigger' struggle between humans and nature. The 'human condition', a device by which we are all made equal and undifferentiated culprits, becomes the target, not capitalism or specific classes in it.

3.7 MARXISM AND LIBERATION

Liberation theory

Heilbroner (1980) asks whether Marxist socialism has to be inextricably connected with the dogma and totalitarianism of China, Cuba, Cambodia and east of the old 'iron curtain'. He thinks that the answer is partly 'yes'. For if Marxism is about uncovering the *hidden* essences of things (see above) then it follows that those who have gained 'insight' may become impatient with those who have not. This produces a tendency to dogma, and, as we will see, utopianism – both of which may facilitate totalitarianism.

Yet the socialist vision inherent in Marxism implies the *antithesis* of totalitarianism. Indeed, largely through common ownership of the means of production and the consequent abolition of private property relationships and classes, people should be considerably freer than they are today, in so-called 'free' liberal democracies.

Many Western socialists insist that although the 'communists' of China,

Russia and the like may have (once) eschewed capitalist production, capitalistic *productive relations* have really grown there. The state became capitalist entrepreneur, keeping real power (that of controlling the means of production) from the people, and forming a ruling class – the Party elite.

However, in 'true' communism people would escape from the tyrannical mind set that now sees capitalistic productive relations as 'natural' and regards seeking alternatives as 'perverse and idle' and capitalism as 'cosmic and unalterable' (Seabrook 1985, 29). Freed from false consciousness, people will change the repressive socio-economic order that *they*, in their alienation, have created or tolerated. They can then *make their own history*: the society which they want, and this will create true consciousness. (The circularity of such reasoning is the perpetual dilemma of socialism, as Orwell's *1984* emphasised. His society could not rebel until it had attained consciousness, but it could not have consciousness until it had rebelled.)

Grundmann (1991, 232–40) spells out the characteristics of communism according to Marx, distinguishing between what he calls 'weak' (socialism) and 'strong' communism – a distinction that, we have seen, not all socialists (e.g. the SPGB) would make. The first involves abolishing private property, classes and therefore class oppression; universalising spiritual happiness and adequate material wealth, and expanding people's 'disposable time'. Strong communism additionally involves returning to use-value production (and therefore abolishing exchange, money and wage–labour), and non-alienation – 'reappropriation of people's objectifications' (including nature, therefore regulating society-nature interactions rationally via conscious communal control).

So, in liberation, society will not be *determined* by 'external' laws: of nature, economics, or history – nor by any utopian blueprint laid down in a previous age.

> Freedom in this field can only consist in socialised man, the associated producers, rationally regulating their interchange with nature: bringing it under their common control instead of being ruled by it as the blind forces of nature: and achieving this with the least expenditure of energy.
>
> (*Capital*, III, 820)

But the release will not be absolute: first, communistic society must recognise ultimate natural boundaries. And, second, we cannot simply do as thought inclines us, uninfluenced by our own historical circumstances. 'Primitive' societies, with limited technological capacity, were much governed by natural limits. Medieval societies were materially and spiritually constrained by 'God's law'. To us today this may appear as ideological mystification, but even to visionaries in these societies such constraints were real enough. And even if wishful thinkers among us today seek to escape capitalist economic laws by creating alternative communities with their own communistic exchange, we cannot totally renounce bourgeois attitudes – inculcated through generations – which make us pine for privacy, individual wealth and the like (Pepper 1991).

119

So Marxism preaches liberation, but not, idealistically, total free will. How a society can organise, at any given stage, to produce and distribute wealth is not a *totally* free choice. It depends on the material state of the forces of production (natural features and resources, available technology and people's productive skills). And what people think and do is not totally open either, but is circumscribed by the material circumstances of history. At the same time, says Parsons (1977, 36) Marxism is against

> religious idealism, against the doctrines of fatalism and predestination.
> . . . Marx and Engels repeatedly called attention to the unique power of
> man to . . . 'act upon nature and society and political institutions and to
> change them'.

However this is not freedom *from* the material world in the sense of detachment from it, but freedom in and through understanding how it works and transforming it with, not against, the grain of physical laws. Unlike animals we can imagine the outcome of our actions and realise purpose through them. We, alone among species, need least to adapt to our environment, for we can change it.

Parsons also insists that Marxism rejects vulgar materialism. Some Marxists, certainly, have rejected it, in different ways. First, as with Habermas, there is rejection of the idea that valid knowledge and thought is simply or mainly *instrumental* (a rational means towards the narrow material end of technological control of objects), or 'practical' (for the strict purpose of communicating messages by embedding them in social norms and everyday interactions so that they reinforce the material status quo). Knowledge should in fact be liberatory: aiming to unmask the ideological interests which really lie behind 'objective' thought and the deformation of everyday language. Such knowledge, in the form of 'critical theory' is an important part of the socialist project of emancipation. Its goal is a universal ideology of undistorted communication for all (i.e. 'true' consciousness) (Kearney 1986).

Second, Bloch, in his 'Hegelian Marxism', departed from vulgar materialism by stressing the importance of ideas: especially the utopian visions which Marx tended to reject. Bloch thought that they create a theology of hope which is the basis for *action*: ideals to act for. This interaction of idealism and material action was, he thought, the dialectic which produces history. History is a function of the subjective possibilities of human consciousness (expressed in various manifestations of the 'superstructure') and the real objective possibilities of nature as the material basis of human change (Kearney 1986).

Unfortunately, socialist utopianism has sometimes been carried to grave extremes which regard any notion that socialist free will could be limited by real external material constraints, including natural limits, as 'counter-revolutionary'. Russian communists were (understandably) so anxious to play up human ability to change society in the 1930s that this led to Stalin's disastrous dictum that nature affected only the *speed* of human development but not its direction, and that communists were the 'masters' of nature. The practical results of this lay in bad farming practices in the interests of maximising production on the collectives, crop failure and famine.

This aversion to environmental determinism was so strong as to insist that Soviet science must be used in support of the prevailing ideology. Hence, Soviet geographers who committed the sin of geographical deviation (environmental determinism) in the 1930s, and even as recently as the 1960s and 1970s, were attacked by the Party. Determinism was declared a scientific weapon of the bourgeoisie (Matley 1966, 1982).

A ludicrous example of this was the rejection of any aspect of Darwinism. During 1930–60 the science of genetics in Russia was rewritten under the direction of T. D. Lysenko. He falsified data, imprisoned Soviet geneticists and set back Russian agriculture by insisting that hereditary changes in plants were not a function of chance, but could be *induced* by exposure to controlled environments. This was Lamarckism (Oldroyd 1980). It was incorrect, and so attempts to breed plants which would have helped the effort to become self-sufficient in food failed miserably, as did so many collective farms. But Lamarckist evolution had been ideologically acceptable, whereas Darwinist evolution, which minimised the importance of socialist willpower, was not (Medvedev 1969, 1979).

Socialism, however, has an ambiguous relationship with environmental determinism. For if environment (as distinct from genes) does determine society, then society can determine *itself*, because society can shape its own environment, as utopian socialists like Robert Owen and Feargus O'Connor have always thought (Hardy 1979). Hence

> The emancipatory power of landscape – or environment as it later came to be called – has always been a powerful, somewhat mystical impulse in socialistic thought.
>
> (Seabrook 1985, 41)

Liberating productive forces and human labour

Marxists may believe that capitalism has enabled productive forces to develop enormously, freeing many societies from close bondage to 'natural laws' (from the realm of necessity to that of freedom). But capitalism has inherent contradictions (see above) which prevent the full fruits of cooperative (social) labour from being further realised. It hinders 'what Bacon envisaged: true human freedom' (Parsons 1977, 16).

Marxism envisages freeing productive forces by first turning them against capitalism, and then developing them in ways that capitalism would not. For instance 'We should try to imagine and build technical regimes compatible with freedom, social justice and other key political ends' (Winner 1986). This would include production under the 'real association' between people which

121

modern machinery could make possible (Mumford 1934, 161–2); open to 'common participation and understanding'. It may entail judging the value of work on other than narrow economic criteria, thus reshaping industrial processes to resurrect creativity for the worker, but at a cost of lower 'productivity' (Schumacher 1980).

Mumford predicted that developing productive forces would create an eventual 'liberation from the machine', as knowledge became more holistic. Thus, as appreciation of our bodies' needs for good nutrition, healthful housing and sounder recreation advances, the need for machine-based sophisticated surgery will diminish (cf. Capra 1982). And older machines will die out and be replaced by faster, brainier ones. Machine-led economic growth will be replaced by economic 'consolidation', i.e. dynamic economic equilibrium. In the society–nature relationship: 'Not mine and move, but stay and cultivate are the watchwords of the new order' (Mumford 1934, 175). One is again struck by how appropriate Mumford's views of sixty years ago are to today's green movement. He mixed ecocentrism with a humanistic technocentrism in a way very appropriate to green socialism.

Mumford was ambiguous about technology as progress, as Chapter 3.6 pointed out. Marx, too, was 'caught between' technology as progress (after J. S. Mill), and as a destroyer of people (after Thomas Carlyle). The resolution was to recognise the historical reality of the latter, and that technology has been up to now a determining factor on development. Yet 'critically', i.e. in terms of what *could* and *should* be, Marx optimistically envisaged an unalienated technology under communism. Through collective social control, it could allow people to 'step aside' from much production and become many-sided personalities rather than appendages to a machine (Grundmann 1991, 127–38).

This last point pre-empts Mumford, and later Gorz, in thinking that production will – and should – increasingly divide into two types. 'As our basic production becomes more impersonal and routinised, our subsidiary production may well become more personal, more experimental and more individualised (Mumford, p. 164). This is the equivalent of Gorz's division between 'heteronomous' and 'autonomous' production, and, like Gorz, Mumford added that this would not be achievable under a craft regime, where division of labour was abolished – thus he rejected an aspect of more romantic socialism sometimes associated, perhaps wrongly, with William Morris.

Gorz (1982) accepts the need for 'heteronomous' (social) production, governed by laws, codes and organising principles, beyond and 'different from' the individual, which would restrain what individuals could do and lead to some loss of creativity and fulfilment for them. Indeed, heteronomous production would involve, in part, the alienation which Marx recognised in capitalist production. But the labour needed by this sphere would decrease over time through mechanisation and increased productivity, and most people would contribute just a little time to it. However, it would never be liberated work in the sense of being self-managed by the producers (Gorz denies (p. 65) that Marx ever equated the idea of liberation with self-management). Instead, liberation would come by virtue of everyone being freed to do 'autonomous' work, i.e. work directly for themselves. This could involve skill, craft, creativity, service to others and self-management.

Gorz's is a 'post-industrial utopian' vision of future work and social organisation which is also 'post-Marxist': rejecting Marx's emphasis on collective, social production. It is riddled with flaws (see Frankel 1987) and is perhaps elitist – envisaging a small group (of intellectuals like Gorz) for whom highly skilled, highly paid, rewarding work in the heteronomous sphere would still be a reality. It is also a very green (and anarchist) vision – reflecting how many green economists (e.g. Robertson 1990, Dauncey 1988) seem to have conceded that *employment* (as distinct from work that may or may not be paid) for all in the form of fulfilling social labour is impossible to achieve: a concession which must delight capitalists.

Mumford and Morris both had more appropriate visions for green socialism. They both thought that no gain in production would justify eliminating 'humane' work, though machines should be employed to banish servile work:

> When we begin to rationalise industry organically . . . with reference to the entire social situation . . . the worker and his education and his environment become quite as important as the commodity he produces.
>
> (Mumford 1934, 163)

Morris, however, would have gone further, to banish commodity production and wage labour, applying faithfully Marxist principles (Coleman 1982a). His view was that work should be voluntary, unpaid, without a 'boss', a pleasure, swappable between people, done with a sense of service to someone and done in groups, for sociability and involving craft (Watkinson 1990). He clearly espoused machines in *News from Nowhere*, to eliminate tedious jobs. Morris thus revoked his earlier, Ruskin-influenced, anti-modern, anti-machine preferences, though he fully embraced Marx's disapproval of the alienated worker as a machine appendage under capitalism. His view of work emphasised use, not exchange value, and production for human *need*:

> Worthy work carries with it the hope of pleasure in rest, the hope of pleasure in using what it makes and . . . in our daily creative skills. All other work but this is worthless . . . mere toiling to live that we may live to toil. . . . Wealth is what nature gives us and what a reasonable man can make out of the gifts of nature for his reasonable use . . . this is wealth. But think, I beseech you, of the product of England, the workshop of the world, and will you not be bewildered, as I am, at the thought of the mass of things which no sane man could desire.
>
> (Morris 1885)

Freeing our communal essence

Although Marxism emphasises the role of class *struggle* in social change it believes in the fundamental sociality of humans, a relatively unchanging aspect of human nature. Indeed, all modes of production, including capitalism, do depend on cooperation. But in capitalism people are alienated from the fruits of their cooperation and therefore their communal essence.

Liberation therefore involves recapturing this essence in 'a future state of truly human communism' where there would not be private property, the town–country division would be resolved, and state, law and classes would be replaced by self-regulating communities of many-sided, all-round women and men, working cooperatively for the common good and expressing themselves creatively (Kamenka 1982a, 12). Marx therefore embraced the Rousseauan–anarchist view of spontaneous, cooperative fellowship as a 'natural' state, given the appropriate mode of production, where people will see society as qualitatively more than just the sum of its individual parts – more than the wills or desires of individual people. The concept of *gemeinschaft*, rather than *gesellschaft* is appropriate, although it is that secular, non-conservative and unhierarchical version of the former, which Marx called *gemeinwesen* (see Kamenka, and Chapter 1.2).

Such ultimate communism, comprising an organic community that elevates (or reinstates) a common life which absorbs individuals fully was to be achieved through revolution. Socialists regarded the short-lived Paris commune of 1871 as a paradigm for it. This involved federalism, decentralisation, participatory democracy, social justice and a rapid improvement in workers' living conditions (Kamenka). Many socialists still see it as their preferred model to that of statism and industrialisation. It is close to anarchist models, but also involves important differences from that ideology (Chapter 5.1). In utopian socialism, the Owenite communes and Fourier's phalansteres embodied some of the commune-ist ideals.

Modern socialist greens often eulogise the Owenite Mondragon cooperatives in Spain (Campbell *et al.* 1977) or the kibbutzim in Israel as the height of communalism, but these have lost much of their socialist–anarchist character. Indeed, enthusiasm for the theory of *communes* (rather than community more generally) as instruments of (green) socialism and social change towards it, needs to be moderated as Marx and Engels themselves maintained in criticising utopian socialism. Visionaries like Sarkar (1983, 168) might imagine that

> By producing as many essentials of life as possible within the commune itself . . . the members would not only assure themselves of a quasi-independent existence . . . they would also, by being able to largely eliminate exchange in the market and by adopting the rotation principle in the allotment of work, be able to abolish alienated work, the commodity character of labour power and the division between mental

and manual labour in some vital areas. They would in this way also be able to develop a rich personality.

But the current reality of communes set within capitalism often comes closer to Gorz's view that communal autarky is impoverishing: 'Self-sufficient' small communes offer their members too few choices in work and lifestyles to attract the majority of people to their ranks (Pepper 1991).

The individual and spiritual

In the face of totalitarian perversions of the communist ideal involving 'vulgar Marxism', neo-Marxists like Habermas have emphasised that the 'universal cannot be realised without the individual' (Kearney 1986, 235). And it must be stressed here that socialist humanism does *not* have to subordinate the individual to the 'tyranny' of the collective. Rather, such humanism should overcome the dichotomy between the two, stressing that fulfilment as an individual can come only through relating fully to the communal group (Marcovic 1990).

Indeed, Marx wanted the full flowering of the powers of the individual, through culture, leisure and creativity for all, via the development of productivity (Cohen 1990). He could advocate this, and yet reject the bourgeois starting point, that individuals are prior to, and imaginable without, society, because he saw the individual as *an ensemble of social relations*. This view has its pitfalls, as Heilbroner (1980, 167) points out, since through it socialist culture will demand 'an awareness of the moral priority of society over the rights of the individual members', and will be 'sharply differentiated' from a capitalist society by being 'as suffused and preoccupied with the idea of a collective moral purpose as is bourgeois society with the idea of individual personal achievement'. It will be a religious society, having something in common with primitive, classical and medieval societies – *and* the ecological societies envisaged by both conservatives (Goldsmith 1988) and liberals (Van der Weyer 1986).

Marxist liberation theory clearly does not stop at material emancipation. The freeing of productive forces is but a precondition for spiritual enrichment in all forms:

> The truly human man, for Marx, must be a total, integrated man, living a total, integrated harmonious life, which is necessarily a fully social life in a community of integrated, harmonious men.
>
> (Kamenka 1982a)

Parsons (1977) cites Marx's view (in *Capital* III, 812) that the more that humans know and control 'the more remote natural consequences of production the more will they "not only feel, but also know, themselves to be one with nature" '. He stresses (pp. 39 and 58) that the word 'feel' is important, for (in *Economic and Political Manuscripts of 1844*) Marx insisted that the aesthetic mode of experience was an essential dimension of human life and fulfilment: part of the goal of human development. Though he never developed his views on subjectivity, he regarded needs for spiritual beauty, love and creativity as 'higher' needs. But they are not separable from lower (material) needs. Marx rejected such a dualism between the material and spiritual.

Indeed, Marx's theory of alienation, and liberation from it, was founded on a deep (liberal) humanism 'in goal and spirit'. The ultimate stage of historical evolution would overcome the estrangements of class society: estrangements from self, from others, from the spiritual, artistic and from nature. As Markovic (1990, 128) puts it, the socialist humanist ethic would 'create social conditions under which all humans would equally be able to bring freely to life their potential creative powers'. Such self-realisation would develop powers of symbolic communication, rationality, creativity, the capacity for choice springing from one's own desires, cultivation of the senses and harmonious social and 'nature' relations.

Thus there is no discontinuity between the creative literary–artistic output of socialists and their socialism. The pre-Raphaelite artists, therefore, for *political* reasons, rejected Raphael as a symbol of the beginning of the rise of capitalist individualism and harked back instead to an imagined medieval world with a deep sense of community (Coleman 1982a, discussion).

Freedom from pollution

Would freedom from pollution form part of true communism? Views about this aspect of socialist production are usually coloured by the debatably relevant issue of what has happened under Russian, Chinese and East European 'communism'. Reflecting the views of the Frankfurt School, Grundmann (1991, 140) claims that 'existing' socialism's pollution record and the global nature of ecological problems illustrate the defectiveness of Marx's technological optimism, for ecological problems are to an extent ahistorical – flowing not specifically from capitalist use, but from technology's inner logic.

Some have claimed that Soviet society in the 1970s took ecological issues seriously (Khozin 1979): a claim which Redclift (1987, 45) calls 'disingenuous', and indeed Elsom's (1992) review of 'communist' pollution bears Redclift out. Others (Parsons 1977, Baritrop 1976) acknowledge 'existing socialism's' dismal pollution record, but point out that it is not systemic within socialism. Rather it shows the gap between such 'socialism' and the real thing: a gap which exists because of the necessary capitalistic phase that aspiring socialism must pass through, or because Russia was (state) capitalist all the time, and 'socialism does not and has not existed anywhere' (Baritrop). Similarly, O'Connor (1989, 106–7) thinks it impossible to say *a priori* whether resource constrained economies [socialist] would deplete resources and pollute more or less than demand-constrained economies [capitalist] ... massive environmental degradation is probably not inherent in socialism as it appears to be in capitalism, although no socialist country has yet demonstrated this proposition.

A more future-oriented perspective acknowledges that a new environmental ethic will require new human relationships, which must be based on a new pattern of production (Thompson 1983). That pattern of production is what decentralist socialists have always advocated (for reasons, mainly, of social justice). So, in *theory*, and if properly applied, socialism would not have to produce a polluting society. For common ownership (which does not *need* to imply undemocratic centralised ownership) would allow resource depletion to be planned, and minimised. And the absence of a market economy would be a *good* thing, permitting full employment, proper wealth distribution, a slowly growing economy and a lack of pressure for consumerism, all of which are inimicable to efficient production.

Indeed, William Morris's (1887) anarchistic socialism would avoid by all means the waste in making wares that do no-one any good, but have involved us toiling, groaning and dying 'in making poison and destruction for our fellow men'. By contrast, *real* needs, beside the material, are

To feel mere life a pleasure; to enjoy ... exercising one's bodily powers; to play, as it were, with sun and wind and rain; to rejoice in satisfying the due bodily appetites of a human animal without fear of degradation or sense of wrong doing ... enjoyment of the natural beauty of the world.

And there must be education, travelling and socially useful, pleasurable work done in pleasant factories or workshops, with pleasant, generous and beautiful material surroundings for all the community.

A hundred years on, this environmental concern suffuses the SPGB (1987) view of socialism as a practical alternative. This 'will have no difficulty' in organising production to 'minimise the negative effects on the environment'. 'Conservation production', with simple maintenance, recycling and few wearing parts, will initially facilitate rapid growth of industries providing quality food, sanitation, health, education, communications and recreation. Then production will fall to a stable level once everyone's basic needs have been met. Information, planning and decision making will gradually devolve from world to local levels, and 'The concept of needs will no longer be based on the idea that increased happiness comes with increased production and possessions'.

Undoubtedly, the key to the ecological benign-ness of 'true' socialism/ communism lies in its economics, which, designed to achieve social justice, should also avoid the ecological contradictions described above (Chapter 3.4).

Predicated on the opposite assumptions to those of neoclassical economics, they hold that there *are* limits to material human needs, and that rational, socially-organised production *can* meet those needs in full for all without destroying people and planet. Rational production (including rational resource use and pollution avoidance), however, cannot come through money or the market (see O'Neil 1988 and Chapter 3.3), or state ownership, but through planning, which is

> indeed central to the idea of socialism [for] socialism is the planned (consciously coordinated) production of useful things to satisfy human needs precisely.
>
> (Buick and Crump 1986, 131)

This will not be an easy task in an economy where people go to distribution centres to take what food, clothing and other articles they need when they need them, and where housing, heating, lighting and water would be supplied free to all, and transport, communications, health, restaurants, laundries, entertainments, parks and museums would also be on open, free access, and where all would work, but voluntarily.

But a rationalised global-to-local network of planned links between users and suppliers can be envisaged, using modern operational research, linear programming and logistic and systems sciences. (Sayer 1992, by contrast, does not think this possible even with the most advanced computer technology – see Chapter 5.2.) The network could hardly produce more wasteful and irrational results than the 'free market' does today, with food mountains in Europe, while Ethiopians starve yet produce coffee and lentils for the European market! Furthermore, in a non-consumerist, stable society, without overproduction–overdemand cycles, 'needs' would be less volatile and more predictable than in capitalism. And where units of calculation need not be expressed universally as money for exchange purposes, the cash nexus will not govern the nature and purpose of economic activity and relationships. Instead other relevant considerations, including the environmental impacts of different products and production processes, can be made significant factors in decisions about what to do and not to do economically (Buick 1987).

Socialist utopianism and liberal distopianism

Marx opposed utopianism in two senses. The first is that which seeks unrealistic routes to social change (Chapter 3.9). The second is that which advocates a static, idealised blueprint of a future society. For the communist good life is a dynamic *process* not a static picture: one of self-realisation and the development and fulfilling of ever richer needs.

> The ideal society, therefore, marks no endpoint in history, but . . . is for ever superseding itself. Self-realisation, then, does not allow for an end-

state called 'communism'.

(Grundmann 1991, 105)

Cohen (1990) regrets that Marx did not specify the nature of institutions in the democratic society he envisaged. This opened the way for 'unfortunate deformities in interpretation' by those who followed. Bloch also thought it undesirable that Marxists lack a utopian vision and indeed criticise utopianism.

But it is a moot point whether Marx and Engels, and even less subsequent Marxists, were *not* utopians in some sense. Coleman (1982a) shows how William Morris refused to look back like Ruskin and Carlyle, those romantic visionaries of an idealised past. Rather, he had a clear strategy for the future, and a view that change is possible. However, this made his socialism essentially a 'vision of utopia [what is desired] which isn't utopian' (i.e. it was not a view that is impossible to achieve). Rather it was 'scientific', precisely because it rejected the 'unscientific' (conservative) notion that new, better, things are impossible because they have not yet happened. Morris thought that if people envisaged a future which was not discordant with nature it *would* come about.

Bloch, too, stressed as an objective factor the world's inexhaustible potential to undergo change. Heaven on earth was a 'potential' space that might indeed be realised when the socio-political liberation of material forces of production coincided with the spiritual liberation of heart and mind (Kearney 1986, 201). This perspective forms the basis of liberation theology in South America today: a Franciscan, holistic, materialistic Catholicism, far removed from orthodoxy. It criticises capitalism, and links material human rights (including social justice) with spiritual values and nature rights.

Its vision of sustainable development is close to socialist utopias; involving material adequacy, high quality of life, accepting technology that increases the efficiency of human labour but desiring to transcend alienated labour, denying limits to growth in science, the humanities and culture, shunning consumerism and nature exploitation, wanting direct democracy (with delegation to higher-than-local bodies) and abolishing power monopolies and bureaucracies (Gudynas 1990). This of course is also close in some respects to green utopianism (Callenbach 1978, 1981).

Goodwin and Taylor (1982) review early utopian socialism: Icarianism, Saint-Simonianism, Fourierism, Owenism, German artisan and early American socialism. None emerged as the direct expression of the modern working class, in the way that liberalism emerged from the growth of the bourgeoisie. Rather, they were the expressions of concern by middle-class intellectuals and leaders, reacting to the plight of workers in early industrial capitalism. And it is as such that Marx and Engels criticised them – as an elite, idealist vanguard rather than part of the spontaneous proletarian material revolution which Marxist purists (like the SPGB) demand. Utopian socialism was thus 'fantastic'. But, say Goodwin and Taylor (p. 162), Marx and Engels's socialism itself had all the elements of the earlier utopians; neither were *they* working

129

class. And despite their attempts to purge their writings of normative blueprints (which is what a utopian vision might imply) they did not completely succeed. Their ultimate vision of social harmony, and an end to 'politics', stemming from the end of class antagonisms, *was* utopian. Thus socialism is a doctrine *for* but not *of* the working class (much the same could be said of ecologism). Hence the difficulty in getting the working classes to accept it.

Liberals have been quick to exploit this, branding (utopian) Marxism as totalitarian, because it depends on what they see as an exclusive and authoritarian outlook emanating from an elite class, but to be imposed on the majority of people. And though Marxism in reality stands against so many features of totalitarianism, it must be accepted that it also stands against some liberal 'freedoms'. By abolishing property and inheritance, through graduated income taxes and centralising credit in state hands it undoes the property basis of liberal 'freedom'. But by socialising production it tries to replace this by freedom from want for all. Hence Marxism seeks to *redefine* freedom (Heilbroner 1980) and in the process eliminates some present freedom.

As Professor John Griffiths wrote in a *Guardian* letter (June 1990) in response to Charter '88's calls for a stronger 'culture of liberty' in Britain:

> What we need is hard-nosed legislation that will drastically *interfere* with the liberties of those who use corporate power to make large profits and promote unemployment, who hold monopolies in public utilities, who prostitute the press, who pollute the environment, who destroy the countryside, who create the poverty of inner cities, who exploit the homeless. Their silly Charter is a trivial irrelevance.

Or, as Kamenka (1982a, 24–5) put it:

> That the contradiction between the individual and the social will simply disappear was said by Marx in his younger days . . . but no-one any longer really believes that. Those who seek an undifferentiated community as the ultimate goal of socialism know that it will have to be created, like the medieval *gemeinschaft*, by force and fraud, by censorship, indoctrination and the ruthless suppression of contrary opinions.

3.8 SOCIAL CHANGE AND META-THEORY

Working-class revolutions

How can and should revolutionary social change happen and be engineered, and by which principal agents and actors? This most vexed of all questions for greens or any radical movement is where orthodox and neo-Marxists seem to be very far apart.

To orthodox Marxists, class struggle is central in liberating humanity from the shackles of capitalism. They await the new millennium when the proletariat, *en masse*, has developed a revolutionary consciousness of the reality of social relations under capitalism and is determined to take united political action to create a new classless society. This will, asserts Parsons (1977, 62), also be an ecological society: 'Labourers, in thought and feeling and action, embrace their fellow human beings and the whole domain of non-human nature as their own'. This scenario, if not inevitable, is strongly suggested by the inherent tendencies and contradictions within capitalism. One tendency is increased immiseration of the proletariat, and another is the increased 'true' consciousness among them which should derive from their being brought close together in towns and factories as capitalism develops. (Even today the relative paucity of agricultural labourers' wages in Britain is said to relate partly to their spatial dispersal, while post-industrial 'utopias' involving a dispersed labour force linked only by computers may really be distopias for that force, where capital has social control.)

Morris's picture of how revolution was to come in *News from Nowhere* (Crump 1990) constitutes one 'authoritative' socialist view of the subject. Critically important was the existence of 'a huge mass of people in sympathy with the movement bound together by a great number of links of small centres with simple institutions'. The decentralisation of these workers' democratic councils meant that they could not all be simultaneously overwhelmed by armed force by the authoritarian conservative government. Also vital was *class* struggle and the general strike. It was not just a movement of 'the people', but of people *as producers*, withdrawing their labour *en masse*, which undid capitalism. Despite the proletariat's willingness to compromise, and its dislike of violence, intransigence and provocation from the far right had rendered accommodation, including parliamentary reforms, a blind alley, and a two-year civil war had resulted.

There was no vanguard of 'great revolutionary men' in Morris's vision: working-class self-education and the *material* experience of political struggle (itself educational) were the catalysts for change. The workers' understanding of communism deepened as they struggled and advanced.

The SPGB today shares Morris's convictions about how revolution must come, except for his contempt for Parliament. Although it dismisses the Fabian route to socialism via parliamentary reform, it holds that revolution through the ballot box is possible – indeed, 'if they won't vote for it, they certainly won't fight for it'. For the state can become a vehicle for emancipation by electing a party (the SPGB) which will enact *one single act* that says 'The earth does not belong to you [the ruling class] but to dthe whole society' (Coleman 1982a). Thus its elected *delegates* (not representatives) will be mandated to put ownership and control in the hands of the whole community, and the need for Parliament and state will fade away. The SPGB, like Morris, stresses that there can be no socialism without a majority of conscious socialists. And it rejects Lenin's argument, in *What is to be Done?* (1902) that the working class cannot achieve that socialist consciousness.

Some Russian Marxists believed in the historical sequence; feudalism, capitalism, socialism, communism: hence they thought in 1900 that the only sort of mass revolution which would take the country out of feudalism would be a bourgeois one (see Chapter 3.2). Therefore if socialism rather than capitalism was to be created a *vanguard* – a cadre of reliable party workers and intellectuals, *for* but not *of* the working class – would have to do it. This view, that Marx was wrong about the revolutionary potential of all the proletariat together, was borne out when, in 1914, Western workers betrayed the revolution by supporting their countries instead of using the war to start international revolution. Therefore Lenin created bolshevism, arguing in *The State and Revolution* (1917) that the dictatorship of the proletariat would have to be state dictatorship (Goodwin and Taylor 1982). (Being used as the legitimator for the infamies of Stalinism, 'Marxism–Leninism', was in fact largely created *after* Lenin's death.)

In the aftermath of Stalinism, many Western socialists today are caught in a dilemma. They want to avoid vanguardism, and may also accept completely the anarchistic argument that socialism

> cannot be created through the conquest of the centres of power [but by] abolition of the conditions of their power, for a society based on relationships of subordination and superposition cannot be overcome by means of a structure of subordination and superposition'.
>
> (Ullrich, in Sarkar 1983, 172)

Hence they are thrown back to the position whereby the working class must

> quantitatively and qualitatively dominate the mass of population and by its own activity achieve an adequate level of class consciousness and an understanding of its historical mission.
>
> (Sekelj 1990)

But this has not actually happened in the West, and many radicals think it increasingly unimaginable. Some, like the SWP (Socialist Workers' Party) feel that this strengthens their case for a revolutionary vanguard. In keeping with *true* Leninism this need not mean what Stalinism produced in Russia. Stalinism '. . . abolished democratic discussion, genuine workers' councils and freedom to argue. [But it] was not an inevitable outcome of leadership . . .' (German, undated, 10). Leadership is not necessarily synonymous with hierarchy, and 'democratic centralisation', based on workers' councils rather than parliament, is not a contradiction in terms, but the logical outcome of genuine democratic discussion. However, once this settles issues, 'everybody, regardless of their position in the debate, has to abide by the decision and act on it' (p. 7).

But others see all class politics as an 'exhausted myth', for the working classes clearly do not *want* anarchism or socialism (Walford 1990). The emergence of a middle class, removed from the productive process, and of a well-paid manual class, has indeed complicated the Marxist scenario. These people, who may have no control over the means of production, nonetheless do not *see* themselves as alienated (Yen 1990a). As Neil Kinnock once told a Labour Party conference, it is difficult to say to a worker with a Volvo, a semi-detached house and annual package holidays: 'Brother, let me save you from all this!' What is at root, think Lash and Urry (1987) is that the material conditions which would produce a politically-conscious working class have disappeared, along with a decline in that class itself, and an expansion of managerial classes. This has accompanied the transition from 'organised' to 'disorganised' capitalism (Fordism to flexible accumulation).

Gorz (1982) has produced a post-Marxist analysis of this transition. Skilled workers – for Marx the major actors in developing proletarian consciousness – have in fact dwindled in numbers with the onset of automation, computerisation and Taylorism. Through the state, and socialisation of production, opportunities for creative autonomous production have diminished. Hence working people have become machines, not conscious of their revolutionary potential, or able to create socialism. And, despite mass unemployment, capital in the 1980s and 1990s has been able to defuse discontent. Partly this is because sites of production are no longer where the decisions are taken. This, and the diminution of autonomous production, has destroyed community accountability, local political life, true democracy and self-reliance. *These*, it follows, are what must come about to achieve revolution. But grass-roots workers' power cannot come in the framework of existing production arrangements, because it is impossible to track down the locus of power in order to conquer it.

Power no longer lies with an identifiable group from whom it can be wrested, but with the system itself. Hence capitalists themselves are largely functionaries of capital, at once oppressors and alienated, not personally answerable for their actions. Revolution cannot come by replacing capital by the proletariat *within* this structure, but by 'collective practice capable of bypassing and superseding [the structure] through the development of an alternative network of relations [i.e. the 'autonomous' sphere]' (ibid., 63). Here Gorz is really substituting an anarchist for a socialist analysis. He has replaced the collective subject of class by the *individual* subject, however aggregated, concluding that collective consciousness is impossible. Only in individual consciousness and autonomy will people find themselves: the working class as a group cannot grasp the organisation of capital's productive and labour system, and therefore cannot transcend it. Indeed, this is a revival of classic liberal theory (Byrne 1985). Gorz has hugely influenced the anarchist strand within the green movement. But he fails to identify how to achieve the collective discipline which will be needed in the inevitable 'outright struggle with the guardians of capitalism' which proto-red–greens like Mumford (1934, 168) foresaw long ago.

Nonetheless many neo- and post-Marxists also abandon the notion of proletarian class consciousness as the agent of revolution, because they see that false consciousness, or 'cognitive dissonance' has triumphed among the proletariat. It started to understand events by reference to the world view of the bourgeoisie during the French revolution, according to Sennett (1978, 237), when the cult of the individual began to displace that of the collective in politics, destroying the working class's sense of itself and its own interests.

So there is widespread proletarian false consciousness, and the locus of the new conflict zone is in ideas, not material conditions:

> the main political problem moves from the economic to cultural spheres
> – how do you struggle against a cultural power which has monopoly over
> the diffusion and production of information and images?
>
> (Touraine 1986)

As Habermas sees it, class conflict is replaced by the struggle between the self-justifying ideology of technological domination, supporting late capitalism, and a critical movement (critical theory) refuting this ideology. The mass media and communications systems, which capital totally dominates, constitute the battleground.

He therefore underscores Gramsci's imperative, of struggle not just in the socio-economic infrastructure, but in the cultural–ideological superstructure, where education, propaganda and the media have manipulated mass consciousness to make it internalise capital's world view. Capitalism stage-managed its own legitimacy, partly by moulding national psychologies, and by creating fatalistic and passive attitudes to those in power. It created an 'ideological hegemony'. The superstructures of civil society are now so adapted to the rule of advanced capitalism that any direct assault on the institutions of state power can be overcome or coopted. So, for Gramsci, and for neo-Marxists (and many greens), any 'war of movement', where proletarians confront the state, must be preceded by a long, slow, reformist 'war of position' to achieve a mass intellectual and moral revolution. In this way human consciousness *can* and does play an active and key role in revolution (Kearney 1986, 169–83).

New agents and actors

So the proletariat are no longer the bringers of liberation, in the view of many neo-Marxists and other radicals. They replace the proletariat by a plethora of groups who, wittingly or no, work for social change. For instance, there are the intellectual dissenters and disenchanted youth who, for Marcuse in the 1960s, had the advantage of 'remaining free from the corporate blessings of advanced industrial society'. But their revolutionary elan was 'either coopted or compromised by the increasingly advanced affluence of the technocratic counter-revolution' (Kearney 1986, 207, 218). No doubt it was also self-defeated by its surfeit of naive idealism, encouraged by Marcuse's own elevation of individual consciousness and lifestyle as a revolutionary agent (see Chapter 3.2).

Marcuse also saw the 'unemployed, unemployable, poor and victims of discrimination' potentially subverting the system; a position which Gorz later embraced in arguing for his new 'non-class of post-industrial neo-proletarians' (Frankel 1987, 209–10). This hardly seems tenable after a Conservative government was re-elected thrice in Britain in the 1980s and 1990s on the back of mass unemployment.

Others (Yen 1990a) put their revolutionary faith in an anarchist counter-culture, because of its opposition to 'all exploitation', such as sexism and racism, not just the economic kind. There is a clear un-Marxist refusal, here, to acknowledge *necessary* relations between such categories.

Yet others (e.g. Schwendter, cited in Sarkar 1983, 173) argue for a wider-than-anarchist counter- or sub-culture, with revolutionary potential: an 'alternative society' distinct from ruling establishment and the class subordinate to it. This 'alternative milieu' includes

> various kinds of communes for production and distribution, residential communes, clubs, study circles, work groups, information centres, journals, health centres, legal aid centres, free alternative schools and 'universities' [etc., and it] . . . has its own alternative norms which anticipate those of the socialist society.

This seems a wide, if not wide-of-the-mark, definition the new revolutionary class, in keeping with postmodernism rather than Marxism, but Touraine widens it still further when he identifies 'the public' as the emerging agents of social change and their enemy as 'the system', consisting of bureaucracy, the state and trades unions.

Thus is the labour movement marginalised by some on the left today. Eco-socialists seem drawn in by this trend, identifying 'new social movements' rather than labour as agents for revolution, and wanting to compromise with their socialist roots by calling for alliances between new social movements and the labour movement (O'Connor 1988). Such calls, regularly echoed at conferences between reds and greens, may, however, gloss over just how much new social movements diverge fundamentally from a Marxist–socialist perspective. Scott (1990) brings out such differences.

New social movements include greens, feminists, and the civil rights and peace movements. They constitute

> a collective actor made up of individuals who understand themselves to have common interests. Unlike political parties or pressure groups they have mass mobilisation or the threat of it as their prime source of social sanction.

And unlike the workers' movement they resist incorporation into institutionalised politics, are anti-authoritarian, and seek value and lifestyle changes

135

rather than traditional political outcomes. They try to create a new culture through personal transformation and new forms of relations. Their challenge to the state is indirect: bypassing it – the anarchist strategy – is often preferred. In fact the aim is to defend rather than demolish the gains of bourgeois revolution and civil society – to defend it from the technocratic state, which is the 'enemy'.

The movements' methodological emphasis is on psycho-social practices (consciousness raising, group therapy and so on), creating free geographical space (urban squatting, rural and urban communes), the 'personal is political' (feminism) and grass-roots democracy (greens). While none of these are incompatible with a socialist approach, the emphasis on their centrality in revolutionary strategy, and on the individual as the locus of revolution, plainly is. Indeed, in the form of New Ageism and deep ecology these strategies can become distinctly counter-revolutionary (see below).

But Scott thinks that Marxist structuralist analysis (such as that by Castells 1978) is often over-reductionist. Conflict, he thinks, is not always reducible to struggle over control of the means of *production*. It is, today, more of a struggle to *consume* more, and better, housing, schooling, health, amenity and material goods, and new social movements acknowledge this. They emphasise a consumer, not a producer, revolution. This ideology, therefore, sees radical (not necessarily leftward) change stemming from ideologically diverse groups, working partly independently and partly through 'networks' (the alternative movement's panacea). They work towards some new social consensus, loosely located around human and nature rights and quality of life. The movements are idealistic and superstructural. They have more to do with ahistorical postmodernism than with Marxism's historical materialism.

Postmodernism

That new social movement theory rejects rather than develops Marxism is underlined by Ignatieff's (1986) assessment of it as part of postmodernist politics. Whereas modernist politics – communist, capitalist or socialist – counselled criticising and acting in the spheres of class and economics, faith in such 'old' politics is dying. Postmodernism tends to place all social conflict in the cultural, not political domain. Its struggle is not to *control* state bureaucracy, but *against* the state. The crisis for conventional politics lies here in the idea that *the public sector* is a synonym for everything cumbersome, inefficient and oppressive, while *the private sector* signifies everything liberating, efficient and responsive.

As Dobson (1990, 157) and Harvey (1990, 46) point out, it is not just greens but many on the left who embrace postmodern politics. These suspect not only the idea of the working class as social-change agents but the whole concept of major agents in universal change. For *universalist* politics end in violence. There are unlimited models of political order, each generated by a relatively autonomous and localised set of practices. So postmodernism sees political problems as *relative*: different for different groups, and this 'disqualifies the search for totalising truth or universalising political ambitions', for the perspectives and approaches of each group are equally valid.

In other words, postmodern politics, like the broader postmodern movement, reject the notions of overarching 'meta-theory' or universally true assumptions behind all theories, and 'meta-narratives', i.e. broadly interpretive schemes. Gorz (1982, 73–4) suggests that 'We are not going anywhere any more: history has no meaning and nothing is to be hoped for it. We can no longer give ourselves to a transcendent cause'. Of course, Marxism involves meta-narrative in its view of history, economics and social change.

Cosgrove (1990) shows how postmodernism in fact recaptures many elements of pre-modern times, including a deep-*felt* sense of moral order in nature, and an *experienced* unity with it (it values subjective, experiential knowledge). Its most powerful environmental icon is the image of the world that was sent back by Apollo 17. To Marxists, of course, this is a false icon, like most deep ecology images of 'one world'. This overworked aphorism (e.g. British television's 'One World' series in 1990 and 1992) may describe the ideal future state, of communism, but to suggest that it is the underlying state of the world today greatly misleads. For to achieve future unity requires recognising today's 'two-or-more-worlds' (that of those who own and control wealth and that of those who do not) as the underlying reality, and the basis for political action.

Some commentators (Frankel 1987, for instance) brand postmodernism as nihilistic, cynical and lacking moral direction. It poses a problem for many greens who, despite the postmodernity of their politics, affirm the need for a clear moral meta-theory emanating from the 'natural order' (Schumacher 1973). This problem may be resolved through envisaging a grotesque society where

> The only things which will not be allowed will be those which do not fit in with the ecological imperative. People will be permitted, but not encouraged, to bugger their children if they want to . . . . you can't repress people on some beautiful idea of what morality happens to be.
>
> (Green Party member, cited in Pepper 1991, 131)

Many Marxists would respond to this attempt to assert the integrity of 'otherness' along the lines of Harvey (1990, 101). Marx's meta-theory, he says, seeks to tear away the fetishistic masks from social relations under capitalism. Postmodernists who proclaim the 'impenetrability of the other' as their creed are overtly denying that there is such a mask. They are therefore complying with the fact of fetishism and being indifferent to underlying social meanings. For if we cannot know or understand how particular social relations sustain a particular moral situation, then it is not necessary, possible, or desirable to be involved ourselves in trying to change those relations. We cannot and should not criticise others or seek to change things – 'anything goes'. But Marxists, being committed to socialism, cannot condone this attitude.

Postmodernism accompanies post-industrial theory, which (deterministically) holds that, largely because of the information technology revolution, Fordist production is disappearing – to produce a regime of flexible accumulation, with relations of production different from those which Marx analysed. And in the new decentralised, culturally variegated, classless but capitalist utopia of this post-industrial society, dominated by service industries, control of information and knowledge rather than the traditional means of production in manufacturing will be the key to political power.

There are flaws in such theories of capitalist utopia (Webster and Robins 1981). But many on the left seem sufficiently convinced that they depict today's reality, and that because of an apparent disappearance of a working-class proletariat in the West, Marxist socialism needs to accommodate to postmodernism. Indeed its perspectives

> pose a challenge to the left's ambition to change the world because they question the belief in rationality and progress which direct and underpin the left's project(s).
>
> (Hebdige 1989)

However, revisionism has in many cases repudiated the very bases of Marxist socialism, as Atkinson (1991, 33) points out in reviewing Marxism and political ecology. Gorz's (1982) view of a dual economy, for instance:

> is not Marxist at all . . . but a very clear reinterpretation of the French philosophical tradition from Descartes to Sartre – that life involves a struggle between an irreducible human essence and the dead machinery of the world around us.

Bahro and Bookchin, other greens from the Marxist intellectual tradition, also have deserted Marxism. Bahro is steeped in New Age idealism and spiritualism. Bookchin replaces the concepts of class and exploitation by 'hierarchy and domination' as ahistoric universals. And political ecology, according to Atkinson, also ought to renounce meta-theory and be postmodern in the sense of rejecting the Enlightenment Project altogether.

Defending the old politics

But others on the left consider that the need for 'old' (i.e. modern) politics is undiminished. Ignatieff (1986) warns:

> Nothing in the end is more dangerous than disillusion with the political process itself. For if we cease to believe that we can master change through politics, we lose faith in the promise of modernity itself.

He also describes ecology and feminism as concerns of the educated middle class, which politically marginalises the underclass.

Yet such an underclass still *does* exist, for whom the struggle is at the material base as much as any cultural superstructure: '. . . we still live in a class society, in which the rich have lately been getting relatively richer'. Furthermore 'that democratic control over the economy [which greens argue for] . . . involves a direct challenge to the power of the capitalist class (and in fact implies their disappearance as a class)' (Ryle 1988, 31).

Furthermore, Gorz is incorrect to assert (1982, 48) that the days are gone when workers felt 'able to exercise unmediated power over production and extend it to society as a whole'. Lucas Aerospace shop stewards clearly showed in the 1970s that they felt able to demonstrate this power (Wainwright and Elliot 1982: see Chapter 5.4). Gorz also exaggerates the extent of de-skilling among the 'neo-proletariat' and does not recognise their still strong desire for 'heteronomous', not 'autonomous' work, and their guilt at not being able to get the former.

Frankel believes that the wage–labour proletariat and their families still form a majority in Western capitalism, and questions how Gorz's 'non-class', without class consciousness, in which no-one holds power, could identify the need to overthrow the capitalist class. He claims that this latter still *can* be identified:

> there are very definite, identifiable persons, groups, classes, interests which do the controlling job, which direct the technical, economic, political machine for the society as a whole. They, not their machines, decide on life and death, war and peace.
>
> (Frankel 1987, 293–4, citing Marcuse)

He also accuses (pp. 224–5) '. . . all those activists in new social movements who believe that it is not worth bothering with trade unionists, and that it is better if unions decay and wither away' of '. . . a poor understanding of the power relations in contemporary capitalist societies'. Without the unions the main actors creating a post-industrial (including green) society would be

> existing conservative parties, businesses, churches and other social groupings, and the new alternative social movement organisations. Given the domination of state apparatuses and the material means of production by the capitalist classes and their political allies, this would, most likely, be no contest.

But even if Marxist class politics *were* outdated in the West, globally they are highly relevant, as Chapter 3.5 above has suggested. Redclift (1987, 170–2) demonstrates that the third world lumpenproletariat is struggling for basic environmental requirements: energy, water, food and shelter. This struggle is an 'environmental' one just as Western trades unionism was – and is.

It is a struggle which is developing alongside two types of economic development, according to Peet (1991, 157). In one, 'peripheral Fordism', mechanised production in countries such as South Korea, Mexico and Brazil, requires importing capital goods and skilled assembly products from industrialised core countries and paying for them by exports of unskilled assembly goods. This, however, allows some development of consumer markets and sizeable middle and skilled working classes. In the other, primitive or 'bloody' Taylorism, Western multinationals transfer limited branches of their production, involving high exploitation rates of the workforce, to states like Taiwan, Malaysia, Philippines, China, Singapore and Hong Kong. This creates nineteenth-century conditions 'And it is not long before the working class reacts in the nineteenth-century manner'. Harvey does not necessarily agree, for he points out how the latter more traditional production systems rest on 'artisanal', paternalistic or patriarchal (familial) labour relations, and these make unionisation and traditional 'left politics' very hard to sustain.

But Redclift is sceptical about the potential for revolutionary success among the third world poor and powerless. They are now immiserised to the point where their lives are endangered, and beyond that where they might be capable of developing class consciousness. Consequently, conflict here is latent rather than manifest.

In fact most critics of class-conflict theory forget what Heilbroner (1980) and Miliband (1989) stress, that it is a theory of *latent* conflict – of normally subsurface tensions, produced structurally through conflicts in the mode of production and revealed by dialectical method. It is about a *secret* history of which protagonists are largely unaware. But, they both insist, the protagonists nonetheless still exist, despite the apparent rise of the middle classes. Miliband believes that notwithstanding the complexity of contemporary class structure a dominant corporate elite with substantial control over the economy and communication systems can be discerned, and it is defended by the state. Heilbroner (1980, 131-2) says that if the proletariat is defined as being without direct ownership of the means of production, its ranks have hugely swelled between the 1800s (80 per cent of Americans were self-employed farmers and urban artisans) and the 1970s (90 per cent work for capital). Now:

> The question may be whether immiseration is a phase through which the proletariat will pass. This question may be tested in the backward nations, where a new proletariat, largely created by the disruptive entrance of capitalism, has plunged millions into urban factory life that is immiserated by any standard.

3.9 THE MARXIST CRITIQUE OF ECOCENTRISM

Since Ensensberger (1974) published his influential critique of political ecology, a steady stream of criticism has been directed from a Marxist position against mainstream greens and green anarchists. Its principal concerns can be inferred from what has been written above, but they are summarised here (see also Ashton 1985).

Apoliticism

Postmodernism's failure to acknowledge the importance of the 'old' politics of class conflict, and Marxist objections to this, have been described above. Inasmuch as mainstream and anarchist greens are 'postmodern', what has been written applies to them. Their position, which reflects a lack of perceived affinity with either capital or labour (Abrams and McCulloch 1976) has steadfastly rejected class politics, from Stoneham (1972):

> almost all theories, liberal or Marxist, about the future development of capitalism, imperialism or the 'third world' will become of strictly academic interest when ecological considerations intervene . . .

to Porritt (1984): 'Politicians of left, right and centre are all both parents and prisoners of the current crisis', to Milbrath's (1989) remarkable assertion that socialism and capitalism have more in common than they have differences.

Lowe and Warboys (1978) pointed out that this is a revival of 1950s' and now 1990s' end-of-ideology thesis: an appeal to end politics and old 'squabbles' and to recognise our fundamental unity in the face of a universal threat to all (in the 1950s the bomb, today, the environment). And they show that this, like the one-world myth, is a *conservative*, not an 'apolitical' message, for it ignores the importance of struggle to change the social order.

In fact the 'environmental' threat is *not* equally grave for everyone. As Ensensberger maintained, 150 years ago industrialisation was causing severe environmental problems but no predictions of doom and ecological collapse were made then, for the bourgeoisie were not affected, as they may now be. Yet if they are rich enough, some can still command amenity and a relatively healthy environment. In their fantasies about ultimate ecological collapse, Harrison (1982) and Elton (1989) both make the point that even as the earth chokes and suffocates, some can buy their way out of trouble. Class therefore *is* relevant, and while it is tempting to think that the 1990s' 'new world order' does mean the end of ideology, the reality is of an emerging East–West 'bourgeoisie' uniting to appropriate surplus value from the third world.

Idealism and individualism

Green political naivety is compounded by a stubborn overemphasis on the power of ideas – of value and attitude change and educational enlightenment – especially at the level of the individual. These, greens have consistently asserted, are what drives history and economics:

> The basic solutions involve dramatic and rapid changes in human

attitudes, especially those relating to reproductive behaviour, economic growth technology, the environment and conflict resolution.

(Ehrlich and Ehrlich 1972)

This approach to social change starts from the assumption that one of the main determinants of a society's structure and dynamics is the individual's values, beliefs, attitudes and concerns. In other words, any change in individual values and attitudes will manifest as changes in all aspects of society.

(MacNulty, in Ekins 1986)

As we think, see, love and do, so our economies are. As we change the way we think, see, love and do, so our economies change.

(Dauncey 1988)

The consequence of this idealism is that insufficient political attention is given to the need to instigate changes in the mode of material production, and to how they can be effected in the light of the powerful vested interests in capitalism which will resist them.

Idealistic individualism can breed an apparent selfishness or, at best, a naive conservatism, as is starkly illustrated by the response of New Age communards to questions on how to deal with exploitation and mass unemployment:

Third world people are exploited. They can't control this. But they can control how they *feel*. They can walk around in a victim consciousness or they can walk around enjoying everything. You go to Sri Lanka and you see the kids happy, happy, happy. You go to America and they're not happy as a race. . . . You can be happy living in cardboard boxes.

Don't worry about unemployment. This and other problems will go away when enough people tune in to their spiritual benignness. They must not view poverty as a drudge but as a gift. Some of the best things in life are learned when you are poor.

(cited in Pepper 1991, 110, 167)

Thus, you create your own reality, which means your own oppression. 'From this it follows that if you choose to believe that "I am no longer oppressed" then the oppression is suddenly gone' (Sjoo 1992). Such sentiments illustrate how

the whole New Age package is a mind-bending and soul-destroying enterprise, its main aim being to uphold present power relations . . . in practice, my observation is that radical politics are denounced as the distressed product of victim consciousness, while conservative ideology and practice are accepted without much question. . . . For, in essence, New Age philosophies fit very comfortably with *laissez-faire* individualistic Thatcherite economics. . . . When faced with problems of

142

oppression or hostility, New Agers advise us to 'just let go of your anger
. . . of your emotional attachments to your cause. . . . Do not, for God's
sake, stay with your feelings of anger, distress or whatever, do not
recognise their source and act on them, including fighting for social
change'.

(Francis 1991, 12–13)

And this counter-revolutionary attitude is also inherent in the elevation of the
individual as key actor in social change. Not only does this celebrate the central
tenet in liberalism, it puts the onus for any lack of social or environmental
improvement *on* to the individual. Thus, as Coward (1989) points out,
alternative medicine meshes well with British Conservative government policy
towards people's health. For it suggests that when illness strikes, or improvement fails to come, the *patient* is to blame: for leading an unhealthy lifestyle,
or for failing to exercise sufficient mental will to get better. As Francis puts it:

For most ordinary people, if you got your foot deliberately stomped on
you would speak out your pain and anger. For New Agers . . . the pain
you experienced in your stomped-on foot was your own creation. . . . If
you are suffering from crippling poverty, the solution is simple enough,
because you, and you alone, are responsible for creating your own reality.
. . . It is your victim consciousness that has created the illusion of
poverty, disease, rape and genocide. . . . Your pain is suddenly your
responsibility, nothing to do with external agents. How convenient for
our oppressors!

This displacement of responsibility means that *your* failure to think yourself
better, or better off, or *your* failure to lead a sufficiently pure ecological
lifestyle, can all lead to the ultimate triumph of the bourgeois weapon of *guilt*,
after which disillusionment and withdrawal from the struggle is but a short
step, and the personal ceases to be political any more (Pepper and Hallam
1988).

Ahistoricism

The failure to set issues in a historical materialist context, i.e. in relation to
changing modes of production, is most clearly illustrated in the population–
resources issue. A historical approach to this is outlined in Chapter 3.5. An
ahistorical approach by contrast refuses to ask what are the *symptoms* of
supposed overpopulation, and why those symptoms could not be the outcome
of poverty, unemployment and wealth maldistribution produced by capitalist
productive relations. It continues the neo-Malthusian fixation of early 1970s'
limits-to-growth theorists, and still insists apocalyptically on the 'reality of a
finite planet' and

143

the unsurprising fact that exponential population growth, combined with increasing per capita consumption of resources combined with increasing destruction and exploitation of the natural environment, is unsustainable, is already resulting in calamity and will result in catastrophe sooner rather than later if current trends are not reversed.

(Ekins 1986)

Ahistoricism extends to a predilection for blaming 'soul-destroying, life-destroying industrialism' or 'the industrial paradigm' (Porritt 1984) for the 'crisis', but not specifying its *form*. Does the fault lie in *all* industrial production, or could we, by adopting proper socialist arrangements, produce, transform nature, reap benefits from science and technology and have *growth* in needs satisfaction and in life quality: all without bringing on ecological crisis? Socialists unequivocally say 'yes': greens are frequently equivocal, vague or just confused.

And there is often ahistoricism over technology. The 1970s' environmentalist messages about technology have been repeated in the 1980s, say Goldman and O'Connor (1988, 92), and they slight or ignore 'the problem of technology as the content and context of social domination, exploitation of labour and accumulation of capital'. Standard environmental texts on technology may assess it simply in terms of cost/risk–benefit analysis, or may see technological harm as a result of lack of appropriate values (Schumacher 1973). Neither approach relates technology to the specific production arrangements or relations of capitalism, or any other mode of production. Neither sees technology, in other words, as an embodiment of specific social relations.

Some ecocentrism may be aware of the vested financial interests behind high technology, but it fails to follow this through by confronting the issue of how to reduce the power which big corporations have over it.

Even radicals like Commoner, Carson and van den Bosch, while relating, 'As in most liberal thought, the economic function of "bad" technology' to 'increased profits' and simple greed and vested interests, do not realise that these explanations are too simple. The economic and social structure of capitalism, and how specific technologies serve the central aim of dominating and exploiting labour, must be addressed. Albury and Schwartz (1982) do in fact do this, showing how technologies as disparate as the Davy miners' lamp, green revolution agriculture, and information technology were developed *specifically* to serve this exploitative aim. Neither they, nor any other technologies, are neutral, and the key issue is not just who controls them. The non-exploitative technology of ecological socialism would be a *different* technology from that of the capitalist mode of production. It is not just liberals, but many on the left who fail to grasp this.

Making a similar point, Winner (1986, 64–6) accuses the alternative technology movement of mere 'sociological tinkering' rather than confronting

144

capitalism. Thus *Whole Earth Catalog* was about technology, and avoided technological politics. Winner's review of 'New Age' writers who focused centrally on technology – Marcuse, Mumford, Roszak, Goodman, Ellul – shows that they identified the problems as human aggressiveness, the machine mentality, the subject–object split, the obsession with technique, rational thought or the second law of thermodynamics. They did not face squarely, he says, the facts of organised political and social power, or examine technology's history (as did Albury and Schwartz) to find out why some choices were made rather than others.

Shying away from underlying socio-economic structures to concentrate instead on the surface manifestations of such structures: this postmodern weakness appears to extend from alternative technology to questions of scale and regionalism. Much ecocentrism identifies spatial reorganisation into small-scale communities (Sale 1985) as *the* key to an ecological society. Most geographers, however, would testify that spatial form *reflects* socio-economic structure, not the other way around. Hence any change to the former could not be sustained without accompanying or prior radical change in the latter.

But bioregionalism (Chapter 4.5), in postmodern style, mistakes the surface manifestation for the structural reality. Thus, one of its gurus, Leopold Kohr (1957, 145) claimed that economic decline was the result of no particular economic system but of *size*. Production units and markets are too big, and business cycles 'result from overgrowth' rather than the converse. Marx's only error was to attribute 'to the system of capitalism what was solely due to the overgrowth of its institutions' (p. 155). Modern disciples reiterate this view of size as a causal factor, advocating a green economics whose '. . . goals should be the decentralisation and democratisation of money and banking' (Weston 1992), not their abolition.

Common ownership and the state

It follows that, for Marxists, ecological improvement must be clearly related to a non-capitalist society. They advocate not merely a redistribution of wealth but a *socialist* society, consisting of a free association of producers. This, says Hulsberg (1985) is what distinguishes eco-socialists from eco-libertarians: the latter have no fixed image of the economics of their desired society. 'Industrialism' is the problem: who owns the means of production is a secondary issue for them.

Green economics certainly are eclectic, as Ekins's (1986) selection from *The Other Economic Summit* papers shows. Sometimes they share the socialist notion of production strictly for need, which figures strongly in Lucas and other cooperative worker–ownership plans. But they also have a strong capitalistic streak: To the Green Party, the 'distinction between ownership and control is regarded as central', but at the same time the party thinks that 'transfer of control should not directly affect the profits made by an investor', and 'rather than pressing for transfer of ownership' it wants to find ways of 'empowering different stakeholders' in an enterprise: workers, community, environment, investors, consumers (Wingrove 1991).

By contrast, a socialist approach should involve common ownership of the means of production, as in William Morris's non-state communism. Its economics are without money or coercive guilt. They enshrine the principle: 'from each according to his abilities, to each according to his needs', whereby everyone has free access to goods and services according to self-defined needs. Production is rational: that is, for need rather than the dictates of any market. Many needs will be met locally, but self-sufficiency is neither attainable nor desirable in the steady-state socialist world society of interlocking self-reliant regions. This society is predicated on Morris's view that abundance is possible because most people's needs are modest if they are self-defined rather than shaped by the pressures of a materialist society (Buick 1990).

Some modern socialists underscore Marxism's conception of a society with no buying, selling, taxation, profits or wages (Hardy 1977). Instead of seeking a 'fair day's work for a fair day's pay' they require abolition of the wages system; this being the only way to end appropriation of surplus value, and alienation in production. Neither can unnecessary trading between nations be countenanced. Indeed there should be no nation states at all, but a series of communities in their regions (Coleman 1982a). In the light of all this it is unsurprising that these socialists should scorn green spokespeople who reveal that they have no particular aversion to the idea of profit (SPGB 1983).

While socialists share much of their vision, together with a distaste for centralised state control, with green anarchists, some are nonetheless impatient with green views that see no role at all for an enabling and locally oriented state. As Ryle (1988) demonstrates, many features of a green economy, like the basic income scheme, could not be administered without some state machinery. And for Johnston (1989) a state is necessarily the medium whereby communality is translated into the planned collective action needed to undo the environmental ravages caused by an unplanned market economy.

Even for Gorz (1982, 112), who wants the transition to post-industrial communism to be direct, without a 'socialist' stage, a state is self-evidently needed '. . . to codify objective necessities in the form of law and to assure its implementation'. In communism the state frees people 'from tasks which they could only undertake at the price of impairing both individual and social relations'. Gorz, like many greens, concludes that we must reject complete state responsibility for the individual, but unlike many greens he also rejects the idea that each individual must internalise responsibility for the material necessities of society.

Anti-humanism

Bookchin's (1992) eloquent neo-Marxian scorn abhors how, in the 1970s, '. . . large parts of the ecology movement began to withdraw from social concerns to spiritual ones, many of which were crassly mystical and theistic'. Along with this, writers like Wendell Berry spoke of humans as 'the most pernicious mode of earthly being . . . an affliction of the world, its demonic presence. We are a violation of Earth's most sacred aspects'. Bookchin says that others, like Damann, or James Lovelock, blame the 'ecumenical we' who 'must be faulted for the ills of the world – a mystical "consumer" who greedily demands goodies that "our" overworked corporations are compelled to produce'. Lovelock (1989, 211) indeed says:

> we, not some white-coated devil figure [scientists] buy the cars, drive them and foul the air. We are therefore accountable, personally, for the destruction of the trees by photochemical smog and acid rain.

Here, clearly, is

> The misanthropic strain that runs through the movement in the name of 'biocentricity', antihumanism, Gaian consciousness and neo-Malthusianism [and] threatens to make ecology, in the broad sense of the term, the best candidate we have for a 'dismal science'.
>
> (Bookchin 1992)

Misanthropy is quite explicit in Gaia theory, as Lovelock demonstrates

> It is the health of the planet that matters, not of some individual species of organism. This is where Gaia and the environmental movements which are concerned first with the health of people part company [p. xvii]. . . . Gaia is as out of tune with the broader humanist world as it is with established science. In Gaia we are just another species, neither the owners nor the stewards of this planet. Our future depends much more on a right relationship with Gaia than the never-ending drama of human interest [p. 14]. . . . This vision of a blighted summer day [smog over Dartmoor, induced by cars] somehow encapsulates the conflict between the flabby good intentions of the humanist dream and the awful consequences of its near-realisation [p. 153] . . . our humanist concerns about the poor of the inner cities or of the third world, and our near-obscene obsessions with death, suffering and pain as if these were evil in themselves – these thoughts divert the mind from our gross and excessive domination of the natural world [p. 211].

Lovelock clearly believes that living creatures, including humans, do and must radically change their environment. Indeed, only through this means has an atmosphere that can sustain life been created: for this atmosphere's composition is inherently unstable, being maintained, nonetheless, because of life's presence. But human-induced change, has gone too far for Lovelock (quite why is not made clear), so that the world has been made 'uncomfortable to live in'. Lovelock concludes therefore that the planet will survive but humanity will give way to

> those species that can achieve a new and more comfortable environment . . . a change in regime to one that will be better for life but not necessarily better for us [p. 178].

Leaving aside the inconsistencies of Lovelock's position, it is obvious that he, like McKibben (1990) both make much, from the depths of their country retreats, of their distaste for seething urban humanity and its problems. Indeed, says Lovelock (p. 210), 'city life strengthens and reinforces the heresy of humanism, that narcissistic devotion to human interests alone'.

Like Thoreau's view of Walden pond 'profaned' by a boat this is modern ecocentrism transforming humans into little more than a 'pollutant', which destroys 'wild' or 'traditional' landscapes. For some greens, the end point of such thinking may be a back-to-the-land rural commune, or even the vigilante violence of American *Earth First!* of which McKibben so approves. Often it can cause friends of the earth to become '. . . enemies of man. It is not that they have ever called for euthanasia or abortions – though that will come – but they regard people as a nuisance' (Bermant 1991).

The 'anti-man' stance has been taken literally by some eco-feminist critiques, as Moore (1990) witheringly observes:

> Instead of denying the tenuous link between women and nature, some women have chosen to reinforce it, calling themselves eco-feminists. . . . Their utopian desire for 'a global ecological sisterhood' may obscure differences of class and race as it brings together a bizarre mishmash of mysticism, morality and the more mundane business of everyday activism. At its most banal it simply echoes radical feminism's division of the world into 'all that is good is female, all that is bad is male'. . . . I can think of lots of good reasons for wanting to save the rainforest, but to claim that it is because the trees are your brothers and sisters has to be the most unconvincing.

Anti-humanism also tends to emerge in the bias towards 'hedgerows, butter-flies and bunny rabbits' of green campaigning about the 'environment' and 'nature' – based, as Weston (1986) puts it, on a middle-class interpretation of what these actually are. This does not see nature as socially produced, or 'the environment' as the suburbs and cities in which most people live. It is a separate 'wild' nature, to be visited or defended from visitors or mediated through TV into the images and fantasies of the market place (Seabrook 1986).

Weston corrects this misperception, suggesting that street violence, alienat-ing labour, poor and overcrowded housing, inner-city decay and pollution, unemployment, loss of community and access to services, and dangerous roads, are the most important environmental issues: all of them produced by the economic inequality and poverty endemic to capitalism. As discussed above, this makes socialism, trades unionism and the labour movement central to environmentalism, not peripheral. But few green groups recognise this. Indeed, few go as far as Friends of the Earth in actually having an inner-city campaign and perspective (Elkin and McLaren 1991).

Incoherence, millenarianism and utopianism

It is difficult to sustain many of the above criticisms against all ecocentrics, because, as is illustrated in Chapter 2, ecocentrism is ideologically diverse: indeed on the definitions of this chapter Marxists are entitled to be regarded as 'green'. But many green activists shun ideological discussion:

> Ask green activists now which is the best route to take to change society, and few will give a coherent reply . . . there has as yet been no real discussion of strategy, no analysis of which tactics are either appropriate or really effective. To address these issues is to step out of the routine of green activism which from the inside feels so secure. . . .
>
> (Andrewes 1991)

This is an accurate analysis because the green consensus about what is wrong and about the primacy of the bioethic in an idealised decentralised green future does not of itself amount to that ideological coherence which Dobson (1990) and Bramwell (1989) sometimes appear to claim for ecologism. For when it comes to deciding on *why*, ultimately, these things are wrong, and therefore *what* to do about them, incoherence can come thick and fast. And, despite its claim to 'deepness' much ecologism is really superficial by comparison with Marxism's structural analysis of our society. This fault is compounded by an ahistoricism which claims 'newness' and uniqueness, whereas the reality is of an endless recycling of old notions: Malthusianism, anarchism, monism, medievalism and so on. Indeed the claim to newness and that the ecological millennium is at hand seems to be the movement's main unifier: its rallying call:

> The social movements of the 1960s and 1970s represent the rising culture which is now ready for the passage to the solar age . . . our current social changes are manifestations of a much broader and inevitable cultural transformation. . . .
>
> (Capra 1982)

Since old ideas are being recycled into such millenarianism, it is appropriate to resurrect old criticisms, and salutary to see how apposite they still are. Marx and Engels's critique of utopianism, especially utopian socialism, is applicable to modern ecocentrism.

They praised the utopians for their awareness of society's evils, but attacked them for their diagnoses of causes. These lacked a materialist, historical perspective, and a class analysis. Thinking themselves 'superior to all class antagonisms' they sought to emancipate 'all humanity at once', in pious hope of cooperation between all classes. And their theories lacked a self-conscious revolutionary proletariat. Engels said that their idealism; their 'kingdom of reason' was the 'idealised kingdom of the bourgeoisie', arguing for absolute truths independent of time, space and historical development (*Selected Works*, cited in Goodwin and Taylor 1982, 73–6).

Their experimental communes constituted a 'fantastic standing apart from the contest . . .' and

> although the originators of these systems were, in many respects, revolutionary, their disciples have, in every case, formed mere reactionary sects. . . . They still dream of experimental realisation of their social utopias . . . and to realise all these castles in the air they are compelled to appeal to the feelings and purses of the bourgeois. By degrees they sink into the category of the reactionary conservative socialists.
> (Marx and Engels, *Manifesto of the Communist Party*, part III)

Again, this diagnosis is relevant to some of today's green alternative communards (Pepper 1991), who, like the utopians, by and large have faith in the independent power of moral example but reject much political and especially revolutionary action. However, as Kamenka (1982a) puts it, though the utopian colonies are inspiring and valuable at an early stage they may become irrelevant when the masses are drawn into the revolution. The world cannot be restructured by moral example, or tiny colonies of well-intentioned people who are not members or representatives of the working class. To do them justice, it is fair to say that most communards seem to appreciate this, although they may not express it in the same terms.

The class interests of ecocentrism

If they are not of capital or labour, whose class interests do greens represent? Many socialists have argued that in one way or another they do defend bourgeois interests.

In a direct, simple way, preservationist and not-in-my-backyard environmentalism protects both the landscapes and the values close to the heart of the bourgeois (see Pepper 1980): it is the ruling classes protecting their geographical and ideological territory. And

> Humane societies and conservation groups tend to arise among the wealthy classes and high-salaried or professional persons . . . often the ruling and affluent classes expend great energy and time on the protection of humanised animals rather than the welfare of brutalised children in home or factory or adult workers reduced to the level of

animals. Their concern for animals is a displacement of human concern for their class position.

(Parsons 1977, 47)

More indirectly, parts of the green movement have become counter-revolutionary through not challenging the material basis of our society but becoming an important part of it; conveying the idea that it *can* continue in a very basic way. Thus, consumerism is acceptable if it is 'green' consumerism and part of 'green' capitalism. The anti-statist, anti-collective, people-must-take-responsibility-for-their-own-lives, individualistic ethos meshes well with Thatcherite liberalism even though the intention may have been different, as does the emphasis on people's power as consumers rather than producers. At the same time, green 'radicalism' becomes a useful warning and corrective mechanism for capitalism, allowing it to adjust to its ecological contradictions and to assimilate protest. And, in green feminist consciousness raising, therapy and encounter groups, the educated middle classes can express their angst – their difficulties in dealing with the spiritual problems arising from their consumerism – and relieve their guilt at being affluent in the face of world poverty.

Finally, greens, as a new social movement, can be seen as a largely third-generation displaced working-class group struggling for status and recognition (Scott 1990, 145–7). From relatively privileged and educated, but not over-wealthy, backgrounds, they lack political power and are excluded from political negotiation in Western countries, which are still essentially neo-corporate states. That is, real power still lies mainly with industrialists or trades unions, and is based on a stable, technocentric set of values, around which capital and labour form a working consensus.

Excluded middle-class, professional groups therefore mobilise at grass-roots level to pursue various causes – peace, feminism, civil rights and ecology – feeding also into the new 'classless' politics of liberal democrats and greens. Their predilection for green causes is rooted in ecology's anti-industrialism: it is anti the main actors in present society. By contrast, in the ecotopia which they desire, *they* – academics, scientists, teachers, carers, community activists and planners – would be the most politically influential people. This, then, is a new sub-class, struggling for the political–economic power to match its social position.

4

ANARCHISM AND THE GREEN SOCIETY

4.1 WHAT IS ANARCHISM?

Anarchism in context

If red–greens would like to push ecocentrism towards Marxist analysis, green–greens often remain stubbornly rooted in anarchist principles (see Table 4.1). Most of the latter do not often acknowledge their anarchist roots, for instance those here described as 'mainstream' greens, including the 'deep ecologists' (Gaians). But they are decidedly there: implicit in the *social* ideals and prescriptions of such greens. It is mainly the communitarian and anarchist–feminist ecocentrics who make their anarchism explicit.

Readers will know enough details about ecocentric programmes and visions, so that there is no need to labour the fact that there are many parallels between them and the anarchist principles and visions which this chapter describes. Political commentators such as Dobson (1990) and Atkinson (1991) do not doubt anarchism's relevance to ecologism. Dobson declares that ecologism's programme is not to be achieved through transnational global cooperation, the nation state or the *authoritarian* type of decentralised commune advanced by the likes of Goldsmith, but through modified anarchist – decentralised, communal and *left-liberal* – ideas of democracy. For Atkinson (1991, 63), political ecology, which informs the praxis of 'green utopianism', is basically an anarchist political philosophy.

On the face of it there is an important distinction between anarchism, which is rooted in nineteenth-century concerns about relationships in *society*, and ecologism, whose proponents – even the most notably anarchistic ones like Roszak, Sale and Bookchin – appear to start from an overriding worry about the society-*nature* relationship. For while anarchism does have positions on human and non-human 'nature', it is not primarily a philosophy of nature. Such distinctions may be academic to some, but for socialists they are vital (see Chapter 1).

If, however, one sees ecologism as a *social* movement (as Marxist socialists do), rather than primarily a development in the attitudes of society towards nature, then it is easy to understand anarchism's conjunction with ecocentrism

Table 4.1 Anarchism: for and against

The features of social life which anarchists broadly *favour* include:
1 INDIVIDUALISM or COLLECTIVISM
2 EGALITARIANISM
3 VOLUNTARISM
4 FEDERALISM
5 DECENTRALISM
6 RURALISM
7 ALTRUISIM/MUTUAL AID

The social features which anarchists broadly *oppose* include:
1 CAPITALISM
2 GIANTISM
3 HIERARCHIES
4 CENTRALISM
5 URBANISM
6 SPECIALISM
7 COMPETITIVENESS

From Cook (1990).

as the philosophy of an alienated and relatively powerless section of the middle classes. For anarchism promises to assuage their post-industrial angst over loss of community, of fulfilling work and lifestyles, of participatory democracy and of control and responsibility over their own lives. Into the bargain, anarchism might just bring people closer to nature (though further from the affluence which bred the angst in the first place).

Anarchism is a fluid and perennially shifting set of ideas and practices (Table 4.1) that on the one hand displays 'postmodernist' propensities, but on the other hand stretches back to the first quarter of the nineteenth century or earlier (to the Diggers). The British movement dates from the 1880s, and enjoyed some popularity before World War I through syndicalism; again in the 1930s, influenced by Spanish anarchism; in World War II and 1960s' pacifism; in the 1960s' New Left movement, and in new social movements and non-governmental organisations (anarchists from Kropotkin onwards have cited, for instance, the lifeboat associations as an example of community anarchism). Scott's (1990) definition of the latter emphasises several anarchistic preoccupations: creating free space (squatting, community urban renovation), wanting grass-roots participatory democracy, opposing hierarchies, centralism, the state and giantism, locally based, and with fluid and shifting organisation and membership. Atkinson (1991) outlines political ecology/anarchism's postmodern credentials, as rejecting the Enlightenment's belief in universal and steady progress. Instead, it wants to build a workable society here and now (echoing Digger Winstanley's plea for heaven on *earth*). Its culture, says Atkinson, would replace Enlightenment rationalism by the empathetic, the aesthetic and the hedonistic, i.e. it would encourage a society which allows us to behave much more according to the dictates of our feelings, in spontaneous pursuit of sensual/aesthetic pleasure, rather than according to some preconceived rational plan or set of social mores.

Anarchism is elusive politically. Marxists tend to brand it as extreme liberalism (Bottomore *et al.* 1983), as do some anarchists: 'Most of classical anarchist theory is a branch of liberal individualism' (Jennings 1990). Others, such as Dobson (1990) and Neville (1990), hold it as a development from both liberalism and socialism (wanting freedom, as liberals, and equality, as socialists). Goodwin (1982) sees it as uncompromisingly socialist. And there are also apparent elements of conservatism in it (see types of anarchism below and Chapter 4.2).

Then again, exponents of ecological anarchism, such as 'social ecologists', find that it answers their almost pathological desire to reject altogether the 'traditional politics of left and right' which have 'aggravated the crisis', and to embrace 'A new cultural politics that encompasses bold experiments in democracy, community and cooperation' (Clark 1990a, 1). Roszak (1979) insists that the kind of individualism of the women's and green movements is not a mere extension of liberal egalitarianism. The way it reaches out to specifically anarchist – small, decentralised and participatory – *social forms*, he says, would be rejected by conservatives, liberals and socialists alike. For it wants institutions to be tailored to the person rather than vice-versa, as in traditional politics.

In the light of all this, it is unsurprising that Marxists, such as Baritrop (1975), identify a lack of unified conscience or purpose as anarchism's main problem. Socialists and anarchists, he further maintains, do *not* have lots of common ground as is popularly believed, nor do they seek similar things. We examine this proposition in Chapter 5.1.

Anarchism's main principles

> It does not matter what particular labels are used: conservatism, liberalism, socialism . . . the criterion for anarchists is simple and direct. Do they accept the authority possessed by the state as the essential prerequisite for the maintenance of social order? If so, then they are of the authoritarian camp.

Here, Neville (1990, 5) highlights the fundamental anarchist objection to any power exercised through conventional politics. For such power may create 'liberty' (as defined under specific laws laid down by an authority). But it does not create absolute *freedom* for individuals; of a sort which is to be circumscribed only by the active, voluntary consent of the individuals concerned. Only that freedom is valid. It represents libertarianism, as opposed to the organised compulsion of representative 'democracy', and, regardless of other political beliefs, it is the starting point of anarchism. Anarchism rejects any higher form of rule, authority or government than that which proceeds directly from the governed themselves (i.e. self government):

> anarchism is the doctrine which contends that government is the source of most of our social troubles and that there are viable alternative forms of voluntary organisation.
>
> (Woodcock 1977, 11)

> To be governed is to be watched over, inspected, spied on, directed, legislated at, regulated, docketed, indoctrinated, preached at, controlled, assessed, weighed, censored, ordered about, by men who have neither the right nor the knowledge nor the virtue. To be governed is to be, on the pretext of the general interest, taxed, drilled, held to ransom, exploited, monopolised, extorted, squeezed, hoaxed, robbed, then at the least resistance, at the first word of complaint, to be repressed, fined, abused, annoyed, followed, bullied, beaten, disarmed, garrotted, imprisoned, machine-gunned, judged, condemned, deported, flayed, sold, betrayed, and finally mocked, ridiculed, insulted, dishonoured. That's government, that's its justice, that's its morality!
>
> (Proudhon, cited in Joll 1979, 62)

Given that government is mainly embodied in the state in most of the world, anarchists are therefore strongly *against the state* in any form, believing that it should be abolished immediately.

While some anarchists think that the state has an independent existence and life of its own (see Carter 1989, below), others accept the Marxist view of it as the indispensable agent of capital. So, for them, to oppose the state is to oppose capitalism, at least in its large-scale forms. Some (mainly right-wing) libertarians believe in utmost individual liberty *and* the 'free' market economy (e.g. the Libertarian Party in the USA). But most anarchists are against, or at worst unclear about, capitalism; especially given that it breeds other features which anarchists oppose.

These include 'giantism': large-scale organisations and structures, which are seen to submerge individualism and self-determination, and to further the domineering interests of financial and political elites. Whether as a large organisation or corporation, or physically as a tower block or huge estate, the giant structure is regarded as remote from individual employees or consumers, and as dehumanising in scale: destroying local communities economically and socially. Such structures are typical of not merely Western capitalism, but the old Eastern state capitalism, which many anarchists, along with liberals and conservatives equate with 'socialism'. Hence the paradox that while anarchists may embrace many of the socialistic principles discussed in this book (e.g. egalitarianism), they will often pillory what they take to be 'socialism', i.e. large-scale state-owned bureaucracies.

Anarchist greens like Roszak argue that the large *scale* of modern living results from 'industrialism'. Scale is a problem independent of whether the form of the industrialism which produces it is capitalist or 'socialist'. Giant organisations, both private and state, are also extremely hierarchical, and anarchists hate hierarchical relationships: personal, economic or political. They are seen as structures where power and control is exerted by some people over others, and this is regarded as the root cause of patriarchy, and the basis of all the repressive institutions of 'advanced' societies, including the nuclear family and conventional education. Removing or sidestepping hierarchy therefore becomes the cornerstone of anarchism, as it is a key to much feminism (which equates hierarchies with patriarchy). Greens, too, attack the idea of hierarchy, especially as expressed in attitudes to nature (Hallam and Pepper 1990).

If hierarchical relationships extend beyond capitalism, then the achievement of socialism and abolition of classes is not of itself a remedy. Anarchists think that the tendency to dominate and control others is a wide cultural, rather than merely political–economic, phenomenon. The manifestations in feminism of an anarchistic desire to replace patriarchy with a non-hierarchical, non-sexist society are seen in women's affinity groups, which echoed from the 1960s onwards the affinity groups of 1930s anarchist Spain (Bookchin 1977). The women who, in the 1980s, set up peace camps outside Greenham cruise missile base in Berkshire especially resisted the idea of leadership and led, elected or assumed spokeswomen, or 'specialists' to convey their message of peace and anti-militarism. Their anarchism also extended to a strong sense of mutualism.

Greens, by and large, have not been as conspicuously anarchic in their organisation as in their political philosophy, except, perhaps for those who are part of the communes movement. While Friends of the Earth do have a strong commitment to local group autonomy, local democratically-organised action, and work through the local community, on the other hand both they and Greenpeace make a point of creating and cultivating hierarchies of expertise to combat effectively the technocratic industrial society. And there are unofficial media gurus, like Porritt or Bellamy in Britain, and intellectual gurus – Arne Naess, Fritjof Capra and Paul Ehrlich for example. Both the British and German Green Parties have agonised over the contradiction between their professed anarchistic beliefs and the need for political organisation, with experts, spokespeople and experienced leaders. In Germany a rift was created in the 1980s, when, after an agreed period, the time came for leaders like Petra Kelly to stand down and assume anonymity. Some argued that for them to do so would waste years of effort spent in creating people with charisma and *savoir-faire*, which made them a powerful political influence outside the Green Party, and therefore more able to extend green ideas. It seems that in the 1990s this 'realist' argument won the day among the British and German parties (see Wall 1991) although it led to the virtual collapse of the former in 1992.

Most anarchists, like socialists, regard economic and social relationships where some people dominate others as leading inevitably to exploitation. They are therefore deemed structurally violent. This begs the question of whether they (should be *overcome* by violence. The cult of violence has been a big element in the demonolology of anarchism in popular consciousness. It is true that physically violent and 'anti-social' anarchism – that of 'propaganda by deed' – has been a minority current running through the nineteenth and twentieth centuries. But anarchism more commonly rejects violence. Some (Tolstoyan) anarchists, through 'bypassing the state' (e.g. setting up alternative communities or local 'black' economies) would bypass the confrontation with capitalist 'law and order' which might spawn violence. Others believe in the tactic of the general strike, and that through it they might find themselves attacked by the coercive arms of capital (military and police) and so be forced to defend themselves, but they would not initiate violence. Yet others embrace pacifist anarchism, sometimes known as Ghandhiism. Through non-violent direct action (civil disobedience) Ghandhi developed the *sarvodaya* ('welfare of all') movement in the face of imperialism (see Ekins 1992, 100–111). It advocated a return to a simple village-scale economic system, free of external control and with a fair distribution of the fruits of the villagers' labours.

This theme of decentralised, local, self-rule has strongly re-emerged in the modern third world *bhoodaan*, or land redistribution movement (Cook 1990) and in the environmental movement such as the Chipkos (Haigh 1988, Ekins 1992, 143–4).

Of course, the *sarvodaya* was a rural movement, and the cross-currents between a certain type of romanticism and anarchism (Shelley, Thoreau and Walt Whitman, for instance, were anarchists) have sometimes influenced the latter towards anti-urbanism. Translated into (romantic) green anarchism this can sometimes become strident. Hunt's (undated) green anarchist manifesto brands cities as biologically unhealthy. He contends that fertility rates are lower in them than in the country. And they are medically unhealthy, because mortality and crime rates are (allegedly) higher than in the country. Roszak's (1979) attack is hardly more thoughtful, berating cities for their bigness, ecological damage, centralised bureaucracy and as generators of extravagant consumer appetites. The world economy and culture, he thinks, needs only small doses of cities, therefore we must 'free' the land from them, for only a minority of people want to live in them: 'The city has always been a mad and murderous place to live'.

But this green brand of anarchism overlooks a rich historical tradition of urban anarchism. This stretches back, according to Ward (1990), to Kropotkin's (1902) description of medieval cities as the home of co-jurations, fraternities and self-governing assemblies, and to Ebenezer Howard's (1898) concern to integrate the best features of town and country. Via Patrick Geddes and Lewis Mumford, this concern underscores modern attempts to create social cities as polynuclear networks of communities (in Milton Keynes, the 'City of Trees') and as 'Greentowns' (Wood 1988). Ward also describes unplanned anarchist cities on the fringes of official cities at the core of centralised economies, such as Brasilia, Chandigarh, New Delhi, Canberra, and Washington DC. The squatter belts of African, Asian and Latin American cities are officially seen as breeding grounds for crime and disease, yet:

> Ten years of work in Peruvian *barridas* indicates that such a view is grossly inadequate . . . Instead of chaos and disorganisation, the evidence points to a highly organised invasion of public land in the face of violent police opposition, to internal political organisation with yearly local elections, thousands of people living together with no police protection or public services.
>
> (Mangin and Turner 1969, cited by Ward)

Employment rates, wages, literacy and education levels are all higher than in the slums from which the inhabitants escaped, while crime, juvenile delinquency, prostitution and gambling are rare. Similar conclusions follow for Nairobi, where a third of the population lives in unofficial housing and creates 50,000 'unofficial' jobs.

Purchase (1990, 10) also champions urban anarchism: 'Social[ist] anarchists argue', he says, 'that the most natural and primary unit of social life ought to be the free, independent and self-governing city'. He looks forward to a new 'flourishing of civic awareness' as seen before in Greek or medieval European cities.

This particular issue well illustrates how fundamental are some of the potential rifts between socialism and green anarchism (Chapter 5.1). Nonetheless many anarchists see such apparently deep divisions as resolvable through the nexus of size and scale. Simcock (1991), for instance, dismisses the green maxim that anarchism cannot work in cities', on the grounds that it all depends on their size. A small ancient Greek-city size of 2000–5000 people would allow democracy to thrive; there is nothing intrinsic in appropriately-sized cities which would lead to hierarchy or over-use of energy. Furthermore, some people *like* cities, therefore ecologism should not want to destroy, but to green' them. Here again is the view of scale as *causal* which Marxists criticise.

Whether in city or country, anarchist political organisation has to reflect the principles described above and also those of mutual aid and cooperation rather than competition. Kropotkin set the tone for many subsequent anarchists and greens, who see these last as vital, instinctual elements in social evolution. They tend, by contrast, to reject competitive international trade:

> This constant pressure to extend the area of markets is not a necessary implication of all forms of organised industry. If competition was displaced by combinations of a genuinely cooperative character in which the whole gain of improved economies passed either to the workers in wages, or large bodies of investors in dividends, the expansion of demand in the home markets would be so great as to give full employment to the

productive powers of concentrated capital, and there would be no self-accumulating masses of profit expressing themselves in new credit and demanding external employment.

(Russell 1918, 112)

But not all anarchism necessarily rejects competition. Proudhon's mutualist society, for instance, comprised individuals and associations (worker cooperatives, banks, municipal administration) democratically organised and competing with each other. It looks like a kind of 'market socialism' without the state, a form which many mainstream greens seem to support (see Ekins 1986 and Dauncey 1988).

The political form must also be *self*-managing, and not one by which people are coerced. As Neville (1990, 5) puts it

Another common distortion of anarchism is the misconception that all anarchists advocate the abolition of organisation. What most anarchists would say is that organisation must be freely entered into, without this it simply becomes a coercive structure. The difficulty here is, of course, the ambiguity of the word 'organisation'. Many anarchists automatically associate the word 'organisation' with an apparatus, even a bureaucracy. The word they use for a gathering together is 'commune' – to commune with each other (in things of mutual interest). To organise has too much of the connotation of being organised. One cannot be being communed or communised.

Indeed, communes are a favoured anarchist unit of social organisation, together with, in cities, neighbourhood groups. The cooperative, in small factories or workshops, is the complementary economic unit. And the general meeting or commune/neighbourhood/town meeting or assembly is the political form that matches them. These are all essentially *small-scale* and *locally* based: the only spatial organisation which will theoretically lend itself to all the other anarchist principles.

Federation between these units is deemed to be the way to deal with issues at city, regional, country-wide and world levels: for instance inter-communal trade, transport, water and energy provision. The world, according to Martin Buber, should be a universal community of communities. The difference between a federal body and a state is that the former is to be run from the bottom up, while the latter tends to run from the top down. Delegates would be sent to federal bodies on an *ad-hoc* or rotating basis, and they would be *mandated* by their local groups. Thus they would be strictly limited in how they should vote and speak, rather than being professional 'representatives' like MPs, with wide discretion and great personal power.

Thus voluntary associations of like-minded individuals – affinity groups rooted in the locality – should run society in direct citizen assemblies, taking decisions by consensus or voting, but preferably the former, when groups have gained enough political skill. Athens, founded on slavery, is not the model here, but the New England town assemblies are. Cooperatives also would involve workers' *self*-management, whether they were competing with others through a market or whether, as some anarchists prefer, there is no competition, no wage labour, no private ownership of the means of production, and distribution simply according to need. Clearly such organisation is far more difficult than political organisation under liberal market capitalism or 'communist' state capitalism. It demands (rather than discourages) a politically educated, caring and involved citizenry: hence anarchists who advocate it are often derided as hopelessly idealistic. Baugh (1990, 101), however, believes that creating such political structures of itself initiates a politicising process. Gradually, by continually participating in decisions about their own political and economic lives, people will be educated into the 'enlarged mentality and common conception of the public good' which is required. There is some evidence from existing communes that this could be the case, but it is clouded by the fact that those who join them almost inevitably come from already educated backgrounds (Pepper 1991).

The concept of community figures strongly in most anarchism, except for the highly individualist form advocated by Stirner (see below). Given that anarchism also prioritises the maximisation of individual freedom, people's mentality, outlook and view of the self are important factors in the anarchist psyche. To create lasting anarchist structures, we must see little or no conflict of interests between the individual and the collective.

At the least, this means arguing that for the individual to accept 'laws', 'rules' or 'restrictions', or that to go along with collectively-reached decisions does not infringe individual freedom, because such obligations, duties, decisions and associations are *self*-assumed – freely entered into, and freely renounceable. But more deeply it means that to be a successful anarchist (anarchist–socialist rather than anarchist–liberal), the concept of 'I' must automatically include the concept of 'we' (see Chapter 1).

As Clark (1990b, 10) puts it, to become communal beings we must 'renew' at the 'most personal level, that of the self': the self is 'incomprehensible apart from one's dialectical relationship with other persons'. This concept of self and society is also a socialist one, where maximum realisation of one's self comes about both by expressing individual uniqueness and through being a social person. Like Marxists, anarchists believe that this is achievable only under true communism, which involves maximum *individualisation*, rather than *individualism*, as in liberal society (see Chapter 3, and Atkinson 1991, 162, citing Lukes 1973).

Roszak (1979) presents his particular anarchistic way as a third choice, achievable in what he calls the 'monastic paradigm', which stresses neither the individual nor the collective. Here, he imagines, false polarities like practical and spiritual and personal and convivial have been successfully resolved for thousands of years. In place of the excesses of bourgeois egotistical individualism or collectivist ideologies, he advocates a mystical anarchism which revives, in romantic individualism, the 'mystery of the person' and draws on Goodman, Tolstoy, Buber, Whitman and Thoreau. Here, he says, the ageless tradition of tribal and village communitarianism is the focus of the self, as is the 'family'. But it is the *extended* rather than nuclear family which resolves the tension between anarchist permissiveness and free growth and the need for authority to hold the community together. In it, conviviality becomes the culminating relationship between free and unique people. This tendency for anarchists to draw upon 'traditional society' as their model is further reviewed below.

Types of anarchism (see Table 4.2)

The apparent tension between individualism and collectivism underlies Woodcock's (1975) classification of types of anarchism along a spectrum from individualism to mutualism to collectivism to anarchist communism to anarcho-syndicalism, involving increasing degrees of institutionalised collectivity.

Max Stirner popularised the first in the 1840s, in *The Ego and His Own*, where he

> sets forth as his ideal the egoist, the man who realises himself in conflict
> with the collectivity and with other individuals, who does not shrink
> from the use of any means in the 'war of each against all' . . . [and who]
> may then enter . . . into a 'union of egoists', without rules or regulations,
> for the arrangement of matters of common convenience.
>
> (Woodcock 1975, 87–8)

His approach, which appealed to the likes of Godwin, Shelley, Emerson, Thoreau, Augustus John and Herbert Reid, is often stigmatised as selfish, existentialist, leading to nihilism and eventually solipsism (the belief that nothing but oneself exists) (Baritrop 1975). But Rooum (1987) argues that to an anarchist 'selfishness' is not a bad thing: if everyone followed their naturally selfish inclinations they would treat others well and be part of society, because such behaviour *is* most naturally satisfying to the self.

Proudhon's mutualism is based on association, not under government, but under a social contract. It allows private possession of property, but not exploitation of others, therefore work is to be organised round a system of mutual credits organised through people's banks. Exchange would be based on 'labour cheques' not money. Federations of local communities and industrial associations would be bound by contracts, not laws, so that social relations are *gesellschaft*, based on self-interest bargains between individuals. Indeed Graham (1989) thinks that Proudhon's mutualism was the highest development of liberalism. However, it did reject the Rousseauian concept of contact between citizens and *government* because in it a general obligation to obey replaces free reciprocal relations. And it rejected a capital–labour contract

161

Table 4.2 Types of anarchism and eco-anarchism

Types of anarchism, after Woodcock (1975)

INDIVIDUALISM
Each individual follows own inclinations, but may enter into a 'union of egoists' for convenience.

MUTUALISM
Work organised around mutual credits. Federations of communes and workers' coops based on social contracts.

COLLECTIVISM
Voluntary groups of people or institutions sharing some goods. Individual still has right to enjoy his or her products.

ANARCHIST-COMMUNISM
Voluntary federations of communes, *collectively* owning property. Distribution according to need. From each according to means.

ANARCHO-SYNDICALISM
Associations based on the workplace. Revolutionary trades unions run all production and distribution, alongside community groups.

ANARCHIST-PACIFISM
Non-violent resistance and revolution. Libertarian communities as a peaceful version of 'propaganda by deed'.

Types of eco-anarchism, after Eckersley (1992)

1 SOCIAL ECOLOGY
Emphasises social origins of environmental degradation. Advocates liberatarian, non-hierarchical relationships. Supports bioregionalism, small-scale, decentralised communities, cultural/biological diversity and appropriate technology.

2 ECO-COMMUNALISM
Adopting human communities to ecosystems. Cooperation and harmony with nature and between people, in simple, socially self-determining small communities obeying ecological laws.

   (a) *Monasticism*
     Asceticism and disengagement from sophisticated society.
   (b) *Bioregionalism*
     Geographical areas with common attributes, in which there is intimate 'biotic' community between people and place.

---

because this cannot be between equal parties. Nonetheless anarchists today generally spurn contractual notions of freedom. Bookchin (1982, 320), for example, suggests that they capitulate to bourgeois ideology.

Collectivism, stimulated by Proudhon and then Bakunin, takes mutualism further by envisaging collective rather than individual possession of goods. People group voluntarily into larger social units of a dozen or more, but individuals still have some rights to the fruits of their labour. Collectivism shades into anarchist communism, the type of anarchism which most communitarian ecocentrics advocate implicitly. As well as appearing most synonymous with 'social ecology' it often seems close to 'pure' socialism. It is most associated with Kropotkin, who attacked contract theory on grounds that it was impossible to determine the moral value of anyone's contribution to society in labour. Contract ideology also, he thought, reflected a shopkeeper's mentality. Since he also rejected the labour theory of value, logically he was driven to the abolition of wage labour, and to distribution according to need, not some notion of who was most deserving (Graham 1989).

Anarchist communism was posited on the slogan: 'From each according to his means; to each according to his needs', and it advocated, through Kropotkin, voluntary federations of communes, each with up to 200 families collectively owning property. Kropotkin's vision is further described in Chapter 4.4.

Anarcho-syndicalism, says Cook (1990):

> sometimes termed revolutionary syndicalism, is the most institutio-nalised form of anarchism (and hence closest to conventional forms of socialism, and possibly, therefore, not strictly anarchism at all), being based on the philosophy of 'direct action' via trade unions. This approach grew in France then spread to Spain, Italy and parts of Latin America in the first quarter of the twentieth century. The concept of the 'general strike' is important ['the revolution of folding arms'], with the workers taking over factories and utilities to establish an alternative society run by the unions, but in practice this part of the movement became largely a conventional trade union movement fighting peacefully for members' rights. An International Workingmen's Association was eventually founded to link mutualist and syndicalist ideas and this still exists, with headquarters in Stockholm. Further evidence of the vitality of this part of anarchism was provided in July 1988 when candidates of Spain's anarchist trade union movement [CNT, the National Workers' Confede-ration] topped the poll in elections in the important SEAT plant in Barcelona's Zona Franca.

Greens should take anarchist–syndicalist approaches to social change more seriously, and they are discussed in Chapter 4.6 and Chapter 5.

4.2 ANARCHISM AND THE SOCIETY–NATURE RELATIONSHIP

The natural society

Like liberals, conservatives and many other political groups, anarchists tend to justify their ideology by describing it as 'natural', i.e. in accord with a perceived natural order (the logical problem of doing this is discussed in Chapter 4.3). Doheny (1991), for instance, describes anarchy as a *natural* state, where humans are naturally questioning and seeking independence, equality and self-suffi-

163

ciency. Like all anarchists, he sees conventional 'education' as a de-naturalising process, which robs children of this innate curiosity, sociability, desire to learn, and egalitarian morality, deliberately moulding them into uncurious clones: mere fodder for capitalism. A de-schooled, anarchist education, like that of A. S. Neil's Summerhill, or Michael Duane's Risinghill, would not mould children, but make learning experiential, drawing out what is already there.

This is but one feature of the 'natural society', described at some length by Hunt (undated) and deemed to be ecological. In his version, people in small groups get their subsistence from working for a third of the year. Naturally, people do not want to work, and there is little reason for them to do so: natural societies are hunter-gatherers. Wealth is evenly distributed; law and order comes through peer and family pressure. And people are also naturally healthy, and not prone to mix outside their community. Thus

> The natural society will not be cultured or liberal or advanced or powerful or hardworking or great: it will be warm and well fed; it will be peaceful, healthy, lazy and parochial . . . a grubby sort of utopia, but the more visionary societies will not work.

There will be no need of a state, of experts and management, of 'progress' and economic growth, or religion or international trade. In 'less developed countries many of life's needs – timber, wild fruit, vegetables, game – are free'. So to make underdeveloped countries as rich as ourselves is simple but inconvenient: we need '. . . to get out, to stop buying food and raw materials from them, and to stop selling them our manufactured goods'.

This natural human society is but a part of the whole biotic society; a natural growth from that which existed before humans. It is a second nature, says Bookchin (1990), which comprises a uniquely human culture and technics, but whose creation by humans was 'eminently natural'. So biological and human natures can never be treated as separate, discrete entities, as the term 'social (second) ecology (first)' implies. Nor should first nature be regarded as merely an extension of the second – which is what Bookchin thinks that the Marxist dialectic does – or the second be regarded as merely an extension of the first – what Bookchin charges deep ecology with. The social ecology (Hegelian) dialectic implies *no centricity*, be it bio- or anthropo-. It 'explains with a power beyond that of any conventional wisdom how the organic flow of first into second nature is a reworking of biological into social reality' (p. 209). Here Bookchin differs from Carter's (1989) view that dialectics are merely mystified systems concepts. Dialectics cannot, says Bookchin, ever be subsumed into systems philosophy.

Bookchin also diverges from the common anarchist propensity to reject utopianism. This propensity stems from the emphasis on the *naturalness* of the tendency to form societies. By contrast, as Woodcock (1975, 21) points out, a utopia is a 'rigid mental construction which, successfully imposed, would prove as stultifying as any existing state to the free development of those subjected to it'. The same can be said of other impositional ideologies, like the dictatorship of the proletariat, Rousseau's Social Contract, or the very concept of the rule of law. But Bookchin nonetheless pleads for utopianism, because, to him, dialectics are not merely about explaining how and why things have been, or what they might extrapolate to, they are about *potentiality*: what could and ought to be. An ecological ethics of stewarding nature is what ought to be, therefore we should not fight shy of declaring it as the basis of our ideal, struggled for, society.

Bookchin defends the bioethic as the basis for his natural society. Though linear rationality might suggest many reasons why a bioethic is illogical, or subjectively and even anthropocentrically grounded (see Chapter 5.2), this does not matter, says Bookchin, because dialectical thinking allows his *emotional* call for biocentricity and stewardship to be regarded as an *objective* statement. Bookchin, however, does not go on to explain why other possible subjective judgements about how second should relate to first nature (e.g. instrumentally rather than through the concept of intrinsic value) are not to be regarded as equally 'objective' and valid.

Humans in their place

Bookchin's society–nature dialectic appears a more sophisticated view of that relationship than that of some anarchists. 'Of course, the community, *like every other animal society*, will have a peck order . . .' says Hunt (undated, emphases added). This suggests that *he* does not see human society as unique, and that, for him, the implication of a monistic nature–society relationship is that humans must, like all other species, follow 'natural' ecological laws (the nature-knows-best principle).

Woodcock (1977, 16–18) also detects in anarchists the doctrine of obedience to the natural (rather than the social) law: as part of a wider underlying belief in a modified version of the great chain of being – an idea which stretches from Plato and Plotinus into twentieth-century ecologism (see Lovejoy 1974). This chain was

> a continuity proceeding from the humblest form of life to the Godhead. . . . Everything had its place in the order of being, and if it followed its own nature all would be well. But let any species break the chain by departing from nature, and disaster would ensue.

Woodcock correctly adds: 'It was a doctrine that might appeal to a modern ecologist'. This is because it is monistic.

Monism allows social ecologists to emphasise that the domination and exploitation of nature by society is but a facet of the domination and exploitation of some humans by others. They are part of the same underlying process. This is similar to another ancient idea, that of the microcosm and the macrocosm, in which people's physical bodies and their societies (microcosm) are held to be but an extension of the biosphere and the whole cosmos (macrocosm). What happens in the microcosm is thought to mirror and be greatly influenced by what happens in the macrocosm (this is the basis of astrology). Ely (1990) calls for modern anarchism explicitly to embrace this idea, along with that of animism. Animism holds that life is in everything. It therefore undermines the customary Western dualism that distinguishes sharply between living mind and (inert) matter. This in turn makes redundant all ideas of a *super*-natural mind (a god) apart from, and controlling and shaping, nature and society, or of the human mind controlling and exploiting any part of nature by treating it as an inert passive object – an artefact. If everything is living, and has mind, then everything can and should be self-organising and self-managing. Animism is thus inherently anarchistic and ecological, especially since, in its pre-Aristotelian form, the concept also implied care, empathy and interplay, or co-productivity, between all the living elements of the universe.

The noble savage

The question then arises as to how we know that the anarchist society is 'natural': that it is in harmony with nature. The answer comes back that there is plenty of evidence for it in the beliefs and practices of societies which are apparently closest to nature: 'primitive' or 'traditional' societies. 'Among the least advanced of the food gatherers, the average size of the tribe is between 300 and 400 persons', says Hunt (p. 1), thus justifying the idea that all societies should be so organised. And he adds (p. 3) that 'primitive man is healthy. Urban man is riddled with disease . . .', making it clear whom we should emulate, and that

> If we honestly want peace and laughter, there is no alternative but the natural society. It will be unsophisticated, but you cannot get rid of poverty or war, or unhappiness, without losing your discos and your symphony concerts. They are all offspring of the same tyranny – obedience to the rulers.

'People living in simple societies have creativity and virtue' thinks Marshall (1989). Roszak (1979) is among many anarchistic ecocentrics who believe the same, and that much of this virtue lies in an ecologically balanced lifestyle. Its basis is the clan, tribe or extended family; the form which anthropology suggests is most common. Roszak wants to salvage the extended blood family, because it is a biological entity with the advantage of spontaneous and unconditional loyalty based on trust. This in turn breeds strongly shared morals, including a land ethic, supported by the mythologies and cosmologies of such societies.

Young's (1990) examination of the anthropological evidence from 'traditional' or 'primitive' people in Oceania confirms some of this. Among Polynesians, for instance, taboo and ritual invoke a contract of the living with the dead and constitute ways of teaching environmentally benign lifestyles. The ancestral spirits of many peoples forbid tree felling, or any exploitative or commercial use of nature. Kinship defines social function for people and ensures cooperation. Environmental success and anarchistic mutualism require this strong kinship – between the living, dead and unborn – an ideological consistency in society of the sort which breeds religious fervour and also strong local identity. 'Oceania's examples' says Young (p. 44), 'may suggest the kind of philosophical reorganisation needed on a world scale if environmental management is to succeed in the goal of sustainability'. However, he adds that we cannot go back to pre-industrial technology or land management to attain our anarchistic ecotopia, though we may be able to selectively re-acquire some of their best features. This would ideally require support for the extended family or, curiously, 'though less reliable, the nuclear family', which is 'probably better than none at all' (p. 153).

Young is at pains to stress that millions in poor countries have that spiritual confidence which deep ecologists seek, yet have very different priorities. Particular people (families) come first, then dead and future kin, then other species, and, lastly, ecosystems. However, Young believes that the (conservative) viewpoint of deep ecology, which would see the individual embedded in society, society embedded in nature and all dependent on cosmic forces, absolutely replicates the Maori world view, which has therefore a contribution to make to an ecologically sound perspective. But he rejects Bookchin's social ecology, as a model that is far from the traditional society. Bookchin opposes the patriarchal family, but patriarchy, says Young, has been part of many cultures which have for a long time achieved a high degree of 'harmony' with nature. Furthermore it 'remains to be demonstrated' that the

> society in which interpersonal relationships were transitory and children were a collective responsibility and the old enjoyed no respect from the young nor felt particular responsibilities towards them . . .

(which is what Young believes that Bookchin advocates) would achieve a harmonious relationship with nature. For 'tribal societies are usually the opposite of this, and experiments in this direction have a poor record of survival' (Young, p. 133).

Reactionary tendencies

Bookchin may be thus distant from a deep ecology/conservative view, but other anarchists are not always so. Indeed, the anarchist's basic concept of a natural order – a chain of being – into which humanity (and humans) naturally fit, is a potentially reactionary aspect of its monist perspective, particularly if the chain is held to be hierarchical. It does not have to be seen thus, for it can be argued that in a chain each link (however humble) is vital for the integrity of the whole – but even this interpretation smacks of the conservative notion of *noblesse oblige*, where the higher forms of life have a paternalistic duty to the lower forms. And the chain of being concept is also extremely reminiscent of conservatism's 'organic community regulated through common values and a commitment to a common life' (the words not of a conservative ideologue, but of Clark (1990b), a self-confessed Bookchin disciple). To the left (anarchist or socialist) the idea of regulating a society through common values smacks of what happens in capitalist and authoritarian societies – people are essentially brainwashed and morally coerced, without thinking or questioning, into a sterile and servile conformity of thought.

Again, some social ecologists see their monistic philosophy of society and nature as best accommodated by the Gaia paradigm (Merrill 1990). But as Dobson has disclosed, Gaia is potentially a reactionary concept. It holds that if humans do nothing to mend their ways ecologically this will *not* destroy the system, which will continue, albeit without humans. It follows from this that those with overweening concern for nature need not act to change society.

Bookchin himself has attacked the mystification and deification of nature which Gaia*ists* (followers of the theory rather than its author) often propagate. Yet some of the social ecology rhetoric seems equally mystifying, if not pretentious. For instance, in social ecology's 'organic society', we are told, nature 'cooperates' with the craftsperson in fashioning a reality either present or latent in it. The craftsperson's work does not simply make an inert natural resource into a desired object, it 'discovers the voice of substance' (Ely 1990). From this perspective it would be rather vulgar to talk, as Marx does, of the worker 'appropriating' from nature.

This hardly seems to be the common people's ecology which the name 'social ecology' suggests. Neither does Hunt's 'green anarchism' seem too people-friendly, unless those people are drawn from one's own bioregion or ecocommunity. For to ensure even wealth distribution and to maintain caring and 'order' the community

> should be a totally separate geographical and social entity. If there is much social mixing between the groups, if people work outside the group, it will weaken the community bond . . . xenophobia [morbid dislike of foreigners, OED] is the key to the community's success.
>
> (Hunt. p. 3)

The 'social order' which Hunt, Clark and some other social ecologists desire to create in order to achieve 'shared ecological values' and correspondingly sound social behaviour is part of their idealisation of the 'traditional' society. Such idealisation is a form of romantic conservatism in itself. It is hard to see how it differs from Goldsmith's (1988) constant appeals for a return to 'traditional' societies, by which he once held up the oppressive Indian caste system as ecologically desirable (Goldsmith 1978), or from deep ecology's call to 'the minority tradition' – a confusing conflation of native American cultures, Taoism and 'some Buddhist communities' with the 1930s' Spanish anarchists and the 1871 Paris commune (Devall and Sessions 1985).

Finally, some expressions of social ecology seem unguardedly to embrace the nature-knows-best-therefore-nature-is-a-template-for-society principle. Marshall (1989), for instance, interprets the message of Kropotkin's *Ethics* (1924) as 'nature is the first ethical teacher of man'. Woodcock (1975, 27) is at pains to show how this belief in the importance of natural law could, but should not, lead anarchism to environmental determinism: 'a passive acceptance of inevitable process'. As such it would clearly be reactionary. Bookchin (1990) is therefore right to reject what he calls the idea of a 'law of complementarity', where first nature is projected onto second nature. This is what he thinks some social ecologists have done, but he insists that the *dialectical* model of evolutionary development proposed by his social ecology is not teleological – i.e. involving societies designed from the outside (e.g. by 'nature's laws'). Rather, it draws out whatever is implicit. It also involves, as we have noted, *subjectivity*, i.e. self-development according to the mind and will and capacity for freedom of the individual and the group who are evolving. Bookchin is thus to be distanced from the more reactionary tendencies of other eco-anarchists.

4.3 ANARCHISTS AND HUMAN NATURE

Common anarchist views and their difficulties

There is no one anarchist position on human nature, which most anarchists, like their opponents, try to use to support their particular ideology (see Chapter 1). Godwin and Stirner were diametrically opposed on the matter (Marshall 1989), for instance. Godwin thought that humans were a product of their environment, which could be altered in order to bring out either innate individualism or innate social (benevolent) behaviour. Both traits were there, as was reason. But Stirner argued that reason, benevolence and solidarity were not possible, because human nature was selfish (even love was a kind of selfishness). Society was a Hobbesian battleground, so, paradoxically, the selfish need to survive would entail people making union with each other.

However it is neither of these, but Kropotkin's ideas which are most usually taken to be *the* anarchist view on human nature, and which inform much modern ecocentrism. 'There is no question about cooperation being the underlying principle of nonhuman life forms', says Sale (1985, 81). He argues that the Leakeys' anthropological work in Kenya demonstrates how humanoids were able to survive 3.5 million years ago by applying their *basic* sense of mutual aid, and that cooperation is today genetically encoded in us. 'Survival of the fittest' therefore means survival of the most cooperative. This, in a nutshell, was also Kropotkin's thesis.

Kropotkin: geographer and anarchist

Kropotkin (1902) maintained that while competition between and amongst species does occur, by far the most important influence on animal and also human evolution is the *natural* tendency for individuals to cooperate and help each other. He believed that mutual aid was a pre-human instinct, and a law of nature (and so in fact did Darwin). He was a naturalist and geographer and *Mutual Aid* is a collection of empirical evidence for mutual aid in animals, 'savages', 'barbarians', medieval society and ourselves. It traces human history and animal behaviour, giving evidence contrary to that *picked out* from Darwin by social Darwinists, notably Thomas Huxley, whose essay 'On the struggle for existence in human society' was published in 1888, provoking Kropotkin to respond. Huxley maintained that because of a fundamental Malthusian tendency for population to outstrip resources, competition and struggle for existence would govern the survival of individuals in a population and be the motor of evolution.

Several important points need to be remembered about Kropotkin's work, to place it in historical and ideological context. Like Darwin and Marx, Kropotkin believed in evolution as a form of *progress* in nature and human society. Like Darwin, he was a natural scientist, geographer and biologist, and believed in understanding nature through observation. He also (like social Darwinists) accepted the premiss that what happened in nature could be a model for human society. And he accepted the existence of competition and struggle as evolutionary factors, but he argued that cooperation and mutual aid are at least as important for natural evolution. In fact he asserted that as far as human society is concerned, evolution comes about *primarily* through *cooperation* within species. His work catalogued historical examples purporting to support this. Most importantly, it was consciously part of a political battle against the right and social Darwinism, and for anarchism. Hence, he tended to pick out from his observations, those which supported his ideology. He saw that

- Animals live in *herds* and hunt in *packs*. Contrary to the idea of intra-species competition, they frequently feed the infirm.
- In higher vertebrates *sociability* is an advantage to evolution and improvement, because it leads to enhanced intelligence (language, communication, imitation, experience), which leads to greater development and survivability.
- *Avoiding competition* is natural: migration is a form of competition avoidance.
- Among humans, individualism was not a feature of early history: tribes and extended families proved more efficient in ecological terms through shared use of resources.
- In medieval cities *guilds* were founded, embodying the principles of

anarchism, where people cooperated for their common good and sought to rule themselves.

- Today, although the cult of unbridled individualism has held sway since guilds and federations were taken over by feudal lords and the bourgeoisie, hundreds of millions still live by mutual aid. There are workers' coops and housing coops throughout the world: there is traditional village life: there are the Swiss Cantons (self-governed, having common land): there are trades unions, voluntary associations and many mutual-aid societies.

The book is packed with examples of mutual aid, and Kropotkin asserts that they are not acknowledged in most biology and history books because these are monopolised by those who benefit from asserting that human nature is hierarchical and unequal. So for Kropotkin there is a human nature, and it is cooperative, not competitive. The state, he thinks, is unnatural.

Other anarchists

They have argued much the same. Proudhon, for instance, published *What is Property?* in 1840. He saw an inherently social human nature resulting from cultural development and therefore historical circumstances. It was evidenced in production: a social act where people pool their labour for mutual gain. And he saw naturally social behaviour, where people instinctively have sympathy, love and mutual attraction for each other. Since property militates against these instincts, and against the social act of producing together for mutual benefit, for Proudhon it was unnatural, and constituted theft.

Bakunin argued, in 'God and the state' (1882) that humans are naturally social: there are 'laws of sociability' binding people together. But hierarchy and authority, especially in religion and the church, tend to destroy this sociability. The church and state are social institutions based on authority, and also the

idea of God implies the abdication of human reason and justice; it is the most decisive negation of human liberty, and necessarily ends in the enslavement of mankind, both in theory and practice.

So for Bakunin, church and state are unnatural. Roszak (1979), a modern eco-anarchist, spells out how this abdication of reason and justice occurs. Western culture and economics rely for their morality, he says, on inculcating guilt and feelings of sin as part of 'human nature': we believe that our nature is *sinful*:

buried away in the core of Western conscience there is a festering accumulation of 'sin' that is simply unworthy of serious concern, and should never have been dinned into our children ... this is the shallowest ballast of our moral nature and we let it hound us through a

171

lifetime of 'good behaviour', 'high achievement' and 'respectability' as we try to prove, again and again, that we are pure, nice and lovable.

Roszak thinks that this 'unnatural' feeling of guilt helps to bolster capitalism, through the work ethic, capitalist acquisition and deferred gratification. Hence capitalism, which he thinks is 'toxic to the planet', will have to go before we can revert to our natural state where we are not guilty about what we are.

There are many difficulties with most of these views on human nature, which some anarchists acknowledge. Brown (1988) argues that all the traditional anarchist arguments are flawed because they contain paradoxes. If humans are naturally cooperative, why have they have acted *against their nature* by setting up the state? If they are naturally social, why have they acted against their nature by setting up property? If social, why have they created anti-social religion and the church? If not guilty, why have we set up a system (capitalism) that requires us to feel guilty? If free and non-hierarchical, why have we set up hierarchies? Why do we keep acting against our own nature?

Marshall (1989), too, criticises traditional anarchism. Stirner's views, he says, are false, Godwin's belief in the power of universal rationality and truth is hard to maintain, and Kropotkin's evolutionary naturalism is untenable. Furthermore, there are no moral values in nature – nature is not the template for society. *Humans* create values so that attempts like Bookchin's to ground an objective ethics in nature must fail. Humans are *not* like other species: they are fundamentally social – born into social relations. This is very much a socialist position (Walters 1980). But Marshall also argues, more anarchistically, that each individual is unique and cannot be aggregated.

He wants to abandon the term 'human nature' altogether. It implies a fixed, ahistorical essence. Anyway, the nature of 'human nature' can never be proved; scientific theories on it are a mixture of observed 'facts' and values. Human values are often projected onto nature and then projected back onto human society through the thesis that humans must take 'nature' as their model (this was Marx's objection to Darwin's theory – in 'nature' Darwin in fact simply saw 'his own capitalist society').

Marshall argues that there is no single model of what humans are like. They are systematically unpredictable, so while social circumstances may encourage good or bad sides they can never be used to express the totality of human nature. We are, in Koestler's phrase, 'holons': 'self-regulating systems which display *both* the autonomous properties of wholes and the dependent properties of parts'. This means that we are two things. We are products of history, and can change our nature only by changing history, and we are also all free, because of our consciousness and intentionality. We now examine both of these possibilities. The first is in the tradition of Marx's materialism, though not necessarily to be expressed as deterministically as it is above. The second is a more idealist, existentialist perspective.

Social product, biological product, or both?

One type of anarchist view takes the historical–cultural argument further. Russell expressed it in 1918 in *Roads to Freedom*. He argued for an anarcho-syndicalist (or guild socialist) society: organised by mutual aid based on the workplace. 'Human nature', he said, is shaped mainly by the economic/social/political system under which people live. It is competitive at present because the system under which people live demands, needs and approves of, such behaviour. In other socio-economic systems it is not necessarily in people's 'nature' to be competitive, acquisitive and pugnacious. Some competition might be natural and good, where it leads people to perfect arts, crafts and public services. Other aspects of competition, which harm people, do not have to be regarded as good, or 'natural': 'The evils arising from these . . . can be removed by a better education and a better economic and political system', and common ownership would have a beneficial effect on human nature.

Russell's view collapses the old 'nature versus nurture' dichotomy (see Chapter 1.2). It says that culture shapes 'human nature', and since cultures vary in space or time, then 'human nature' also varies. This can be argued in terms of cultural or economic determinism. In the latter, human nature is a function of the relations of production in a society, which correspond with the mode of production. Consequently violence, competition, hierarchy and greed can all be seen specifically as products of capitalism (this argument does not say that they cannot also be produced in other modes of production).

This kind of determinism is not one where humans have no control: it is not fatalism. For, since the prevailing economic system and culture are *human* creations then humans can change them for other conditions. The same argument could not be extended to a human nature determined by genes. (Though this is a moot point nowadays with the possibilities of genetic engineering, while behaviouralism and behavioural engineering – see Skinner's *Walden II* or Burgess's *Clockwork Orange* – have long been advocated as ways of modifying fundamental 'inherited' characteristics.)

This is a perspective which utopian socialists have embraced. It sees no pre-programmed animal instincts which prevent socialism. The changes in human nature which have happened over 40,000 years are due to cultural development, not biological adaption. And if we have created technology and culture, thus changing ourselves so far, we can change further, into socialists by 'nature' (Walters 1980). Robert Owen's nineteenth-century experiments were based entirely on this proposition: that if you create a nurturing cooperative environment, then you create similar sorts of people. The same reasoning underlay many alternative communities throughout that century (see Hardy 1979).

A more sophisticated Marxist perspective would add that although humans do make 'their own history' (i.e. future), they do not do it independently of their previous circumstances: these limit us. This is why people from bourgeois society, educated and socialised into a liberal world view, cannot shake off that view merely by living in communes. Their attempts to be communists are constantly frustrated by over-individualism (as self-centredness), passion for private space, reluctance to share and to place community needs on a level with their own and so forth. Hence in the fabric of the very opposition to free market liberalism in the 1980s, eco-anarchist communards who form part of the *counter*culture have become more 'Thatcherite' in their views and actions (Pepper 1991, Rigby 1990).

A more sophisticated Marxist perspective would also argue for a dialectic of nature and nurture. The two variables of culture and genes are constantly interacting to shape each other in a close, intimate relationship where it is hard to separate one from another.

Rose, Kamin and Lewontin (1984) take this perspective. They reject both biological and cultural determinism as *complete* explanations. They say instead

> We must insist that a full understanding of the human condition demands an integration of the biological and social in which neither is given primacy . . . over the other but in which they are seen as being related in a dialectical manner.

So, human nature is *simultaneously biologically and socially constructed*. We *do* have limiting characteristics of our biology which affect our relationship with the environment: we are bipedal, one to two metres high, and we do not have wings. But as our cultural development proceeds these restrictions become less limiting. Our 'second nature' enables us to fly, or communicate over long distances so easily that we do not think twice about doing these things. It has become *natural* for us to do them. And even our very basic natural functions – eating, procreating, for instance – are not just 'natural', for each culture imbues them with different social meanings and overtones.

Existentialism

A different solution to the problem of human nature comes from Brown's observation that anarchists are forced to concede that there are numerous instances of people behaving in ways that are not 'natural'. She argues that the only way to overcome this difficulty is to argue, along existentialist lines, that there is *no such thing as human nature*. To put it another way: we are able to determine our own nature for ourselves: through our own individual actions we create our own, always changing, human nature.

> We have the freedom [the free will] to create ourselves. Of course, we may choose not to become anarchists: we may instead choose to be fascists or capitalists. The point is that *we* choose whether we want to or not, whether we acknowledge our choice or not. . . .

She abandons any determinism, cultural, economic or biological and argues that there are no causal mechanisms at work, sophisticated or not, which determine our nature. She says that we have a much more direct control over what we are, all of us as individuals – and this last is important to anarchists. She cites Herbert Read, who argued in 1949 that anarchism and existentialism were essentially connected because they are revolutionary and militant doctrines which emphasise freedom, especially freedom of the individual. Existentialism allows us to create our own meaning for our own lives, free from any ideas that suggest that we must be subordinate to the operation of physical or economic or social laws: 'the human person, you and I . . . and . . . everything else – freedom, love, reason, God – is a contingency depending on the will of the individual'.

Read claimed that this is empirically true. History shows us that humans act in any number of ways: at times we are cruel, brutal and violent, but we also exhibit love towards one another and are altruistic. If we accept this as anarchists, we are also, presumably, prepared to accept the corollary which Sartre confronts us with. If we do have such control, and then behave uncooperatively and oppressively, the responsibility for the results of our behaviour lies squarely with ourselves.

This leads to potential political difficulties and ambiguities for anarchists, as Yen (1990b) points out. The ideology of free will stresses the freedom and importance of the individual, and it is no coincidence that it became prominent in the USA. In such purely interpretive sociology (as in bourgeois psychology) the individual's experience of, and 'responsibility for' his or her own poverty, crime or illness, for example, may be focused on at the expense of general social patterns or causal explanations. In this way anarchists could find themselves lining up with politically reactionary forces, such as the New Agers (see Chapter 3.9), or indeed the Thatcherites, who dwell on 'freedom', 'choice', 'taking responsibility for our own lives', founding cooperatives and housing associations and so forth. As Yen stresses, the right uses free will as a stick with which to beat the poor and unemployed and blame them for their own plight.

The right, indeed, often mouths 'anarchist' slogans but then, critically, adds to them a different, non-libertarian twist, placing a coda onto 'freedom' which says 'within the rule of law', the social contract, the bounds of 'decency' or 'respectability', or the iron 'laws of economics'.

Perhaps, then, anarchists should not opt fully for Brown's resolution, but should instead opt for Marshall's 'soft determinism'. This recognises that there are causes which influence us, whether cultural or biological, but argues that they *dispose* us towards a behaviour without determining it. Thus is the vital anarchist principle made possible, of self-regulation to produce good social behaviour, without coercion. Unfortunately it was on this point that Russell parted company with his earlier anarchist leanings in 1948, after the experience of World War II. Though still wanting guild socialism and the abolition of capitalism, he felt he had to concede (1918, p. 94; 1948 revision)

There were wars before there was capitalism and fighting is habitual among animals . . . man is naturally competitive, acquisitive, and to a greater or less degree, pugnacious.

Respect for the liberty of others is not a natural impulse with most men: envy and love of power lead ordinary human nature to find pleasure in interferences with the lives of others. If all man's actions were wholly unchecked by external authority, we should not obtain a world in which all men would be free. . . . I fear it cannot be said that these bad impulses are *wholly* due to a bad social system, though it must be conceded that the present competitive organisation of society does a great deal to foster the worst elements in human nature.

4.4 THE ANARCHO-COMMUNIST UTOPIA

Utopianism

Both Marxism and anarchism reject the *idea* of utopia for reasons discussed above: it could become a template imposed by present on future generations. It could restrict their freedom by creating a prescribed blueprint for living, and therefore became a basis for totalitarianism. It is also a recipe for political naivety in the present.

Yet both Marxism and anarchism have strongly utopian leanings, as befits their clear commitment to a world which is much better than today's. In Marxism the commitment is unambiguously socialist: in anarchism as a whole it has elements of liberalism and socialism – and even, perhaps, conservatism (where it idealises the past rather than espousing a future utopia).

But, through anarcho-communism, anarchism and decentralist socialism come very close. They both fix on the commune and/or free-governing city and neighbourhood as the basic social/economic units of the ideal society; and on the importance of *community* in all political relationships (there are important differences, too, discussed in Chapter 5.1). Hence their commune-ist utopias – based on *spatial* as well as social form – are very stimulating to the visual imagination. They imply a hugely radical departure from those spatial forms and landscapes of present capitalist society (or state 'communist' societies).

It is instructive to follow the imagination in this direction: to envisage the geography and landscapes of an anarchist society, because in so doing we can most clearly observe the congruence between anarchist–communist and green utopias. Even though they may *start* from different precepts about what is most important, their ultimate vision of what they would like society to be like has many coincidences: compare, for instance, Russell's anarchistic guild socialist utopia, Table 4.5, with the Green Party programme in Kemp and Wall (1990) (especially the basic income scheme and the importance of the informal economy). The ecocentric vision was spelled out by Callenbach (1978,

1981) and Goldsmith *et al.* (1972), in what are now green classics. And in Kropotkin's *Fields, Factories and Workshops Tomorrow* (1899) and Morris's *News from Nowhere* we find anarcho-communist principles translated into landscapes and socio-economic detail that are similar to the green landscapes.

Morris was a socialist rather than anarchist – his Socialist League was often quarrelling with anarchists. Nonetheless, as Woodcock (1975, 21) points out

> the only complete utopian vision that has ever appealed generally to anarchists is *News from Nowhere,* in which William Morris, who came remarkably near to Kropotkin in his ideas, presented a vision – charmingly devoid of any suspicion of compulsion – of the kind of world that might appear if all the anarchist dreams of building harmony on the ruins of authority had the chance to come true.

Woodcock (p. 22) goes on to describe the tension within anarchism which '. . . often seems to float like Mohammed's coffin, suspended between the lodestones of an idealised future and an idealised past'. That tension is illustrated by contrasting Kropotkin's considerable faith in technological progress to achieve high population densities with Morris's utopia, generally bereft of masses of people and distasteful of the sprawling suburbs of London which threatened to engulf the countryside. As Woodcock says (p. 21):

> The middle ages are in fact more real to the inhabitants of Nowhere than the chronologically much nearer nineteenth century. The idea of progress as a necessary good has vanished. . . .

Yet one must beware of aligning Morris, because of this, with romantics like Thomas More, William Blake and John Ruskin rather than utopian socialists like Robert Owen, Charles Fouvier and Henry de Saint Simon, who saw technology as vital to wealth creation and the liberation of the masses (Redmond 1983). For Morris wrote much more than this novel, and the totality of his works made it abundantly clear that he espoused Marxist realism rather than romantic idealism, and looked forward. A future of 'meaningful work' rather than 'meaningless toil' was very much a vision of progress derived from Marx (Coleman 1982a), as was the idea of abolishing the town–country distinction in favour of a more equable population distribution (proposed in the *Communist Manifesto*).

Principles of landscape evolution

From Chapter 4.1 it will be gleaned that the main principles underlying the evolution of landscape and geographical space from capitalist to anarcho-communist forms are: decentralisation and smallness of scale; blurring town–country differences (combining the best of both); self-reliance, locally, regionally and 'nationally'; anti-specialism; provision of abundant socially useful and fulfilling work; and social and economic equality and direct democracy.

Given these, we might expect some or all of the following to be elements of anarchist communist landscapes and geography.

Industry, work and the distribution of settlement

Everyone can enjoy the advantages of the countryside and nature along with the advantages of urban life. Communes and small villages are the common rural form. Cities are reorganised into community neighbourhoods and 'greened'.

Production is where people live, in small factories and workshops in each village and neighbourhood. It is mainly small scale: 'economies of scale' are not sought. Largest units have twenty to fifty workers; smaller units have less than twenty workers. The purpose of production is to create meaningful, fulfilling work, and socially useful goods and services, so *less* will be produced. There is far less division of labour than under capitalism. Individuals do many aspects of production: they do not specialise greatly.

Common ownership of land, buildings and machinery leads to less pollution. For there is nowhere to 'externalise' it to as now, and no incentive to try.

Power generation is in forms and on a scale that ordinary people can own, understand and operate for themselves. This leads to the disappearance of large (especially nuclear) power stations, though not necessarily the removal of a national grid. Insulated, energy-efficient houses and soft-energy-generating equipment are the norm.

People do not rely on money for exchange. They take what they need, when they need it, from what is produced. They share resources. There are no money markets. International trade is much reduced. There is no profit motive. Most regions try to achieve a high degree of economic self-sufficiency.

Central business districts with banking and insurance high-rise offices, airports and container terminals all disappear. There is little road traffic by comparison with rail. 'Shopping centres' become depots for free distribution of goods and services. Street furniture no longer contains advertising. There are large areas set aside for community arts (street theatre for instance), for public meetings, and for children's play.

Government, law and order

Representative parliamentary government, with the state and all its apparatus, are abolished. People meet together locally to decide directly on all policy for their locality. But they may form federations: into regional or international assemblies to deal with matters beyond the locality.

All buildings associated with a central coercive state have disappeared (e.g. inland revenue offices). But there are many buildings associated with local and community self-government, including those big enough for local and regional assemblies. Law courts and police stations have gone, since there is no poverty and therefore little crime. As there is no war, because of no international competition for resources and trade, there are no military airfields, docks, barracks and parade grounds, etc.

Agriculture, nature and beauty

People farm collectively, and not for profit. Most people spend some time working in craft production in workshops and factories, some time in the fields, and some in education.

There is little specialised agriculture. It is labour intensive, but it is also highly scientific, and based on working with and not against nature.

Human sewage is used on the land. The soil is well protected from water and wind erosion. Fields are small, with much hedgerow, coppice, woodland and other wildlife refuges.

The landscape therefore has a 'bocage' appearance, alternating with particularly intensive cultivation around each village, commune or town. Beauty and diversity of landscape – 'wild' and farmed – are highly valued as sources of human wellbeing.

Transport, recreation and leisure

Principles of local production for local needs and minimal energy use, and the reality of satisfied and fulfilled lives in the community all make for less need for transport for material purposes or recreation. There is no tourism, but there is travel to increase knowledge and appreciation of diverse cultures and localities.

People are happy to travel as they live, with other people, in public transport.

Physical fitness is highly valued as part of human fulfilment along with mental development, but simple and unobtrusive forms of getting fit are favoured. Competitive sport is not. People generally make their own amusements.

Large theme parks, sports/leisure complexes, multi-entertainment centres, pleasure parks and the like have disappeared. Spaces are provided everywhere for people to create music, art, games and recreation spontaneously.

Nowhere and Fields, Factories and Workshops

Work, industry and the distribution of settlement

When Morris's hero wakes up in the twenty-first century (having gone to sleep in the nineteenth), his first impressions of the banks of the Thames at Chiswick register an absence of the familiar soapworks with their smoke-vomiting chimneys. There are no engineering shops or lead works, and no sounds of riveting and hammering. We learn that such factories which do exist are called banded workshops, where those who still want to work together in large-scale production (for example, making pottery and glass in big ovens) can do so. But on the whole production is small-scale and for local use rather than for distant and 'artificial' markets.

Kropotkin's view of work and production advocated small workplaces where people already live (i.e. more rurally). They must be inexpensive enough for there to be many of them. Production methods must be simple, minimising the demand for high skills (and therefore, into the bargain, the organisation of production could be more democratic – not revolving around 'expert' elites). And, once again, production must be locally-based for local use. Kropotkin did not eschew using machines to save labour. They were welcome, if small and simple. But handwork would extend its domain, particularly in applying artistic finishes to products. Morris, too, wrote of machinery replacing irksome work, but not the creative work so needed for fulfilling mind and body. In *Nowhere*, the machines have been 'quietly' done away with and handicrafts rediscovered to a far greater extent, one imagines, than in Kropotkin's Britain. For there is much mention of craft workers – weaver, thatcher, printer, boatworker – as well as the administrator and organiser, whose job it is to eliminate waste. Morris's characters do not do just one job, but, in accordance with eliminating over-specialism, they will leave their boat duties to go haymaking, or their weaving to have a break by rowing the ferry. Kropotkin's workers, similarly, spend part of each day in the factories and workshops, and part in the fields, in 'integrated labour'.

To combine work in this way, and also avoid the social and moral excesses of centralised urban-based capitalism – 'masses in misery' in Dickensian squalor – agriculture and industry are reintegrated. Kropotkin and Morris are very close on this. Capitalist industrialisation drew people from the land, and in the resulting cities people forgot the bonds attaching them to the soil; these bonds are to be re-established.

Given all this, and their principle of local production, both writers envisage the 'scattering' of industry over the world, and over the territory of each country. Kropotkin demands a transformation in the relations between labour and capital:

> a thorough remodelling of the whole of our industrial organisation has
> become unavoidable. The industrial nations are bound to revert to
> agriculture, they are compelled to find the best way of combining it with
> industry, and they must do so without loss of time.

He tried to show that in the 1890s, already, most of British industry was in small factories of between twenty and fifty workers, or workshops (defined as without electric or steam power) of less than twenty, and that petty trades and rural industries and crafts abounded. This kind of organisation was natural and desirable, and concentration into large-scale enterprise was not an economic necessity. However, to compete with what large-scale industry did exist, smaller enterprises would need to federate and cooperate.

Hence, Kropotkin's landscapes feature the small factory amidst the fields, where industry had come to the village: not in capitalist form but as community-organised production. This way, the workers regain possession of the soil around them (there is to be a multitude of small landowners, implying a multitude of field boundaries) and they cultivate it.

This scattering gives a very dispersed settlement pattern, as is evident in *Nowhere*. City suburbs 'have melted into the general country', although small towns have not been cleared. (They have, however, been substantially rebuilt, and most have become nearly as beautiful as Oxford was.) People have 'flung themselves' on freed land, and the villages have become more populated than they were in the fourteenth century (reversing the rural depopulation of Morris's day). After the people's revolution, the town has invaded the country – 'the difference between town and country grew less' – but the invaders 'yielded to the influence of their surroundings and became country people', while the world of the country is vivified by the 'thought and briskness of town-bred folk'.

In *Nowhere*'s Britain, it is therefore virtually impossible to be out of sight of scattered country houses. The houses are generally small. Large 'cockney villas' of the type that once lined the banks of the Upper Thames, and were lived in by the rich, are gone. Houses might be occupied by separate families, but the door is not shut to the 'good-tempered person content to live as other housemates do'. And there is some multi-occupancy, symbolically of Windsor Castle. But Fourierist-style 'phalangsteries' are ruled out, for these large units of communal living are seen as a response to poverty, and poverty is now extinct. However, the unit of management of an area is the commune, ward or parish, which is run by meetings that reach decisions by a mix of absolute consensus and majority voting. The meeting house, with the theatre and market (where, as in all 'shops', no such thing as money exchanges hands, and people simply take what they need), form prominent buildings in most villages.

The city: greened, decentralised, or gone

Just as capitalism led to the agglomeration of people and production in industrial cities, anarchism leads to the reverse. Kropotkin envisaged that the city would not last, and Morris's England has duly lost, completely, Manchester and most other cities except London.

> As to the big murky places which were once, as we know, the centres of manufacture, they have, like the brick and mortar desert of London, disappeared: only, since they were centres of nothing but 'manufacture',

and served no purpose but that of the gambling market, they have left less signs of their existence than London.

The elimination of poverty leads, in Morris's mind, to the elimination of slums, which he appears to regard as synonymous with high-density living. That sense of community which we frequently associate with dense (inner city) housing in manufacturing areas of Britain is not acknowledged. For Morris it comes only with proximity to the countryside.

Appalling manufacturing places and practices need no longer be tolerated: 'Whatever coal or mineral we need is brought to grass and sent whither it is needed with as little as possible of dirt, confusion and the distressing of quiet people's lives'. Morris gives no details of how this is to be done: the fact that it *is* done will, however, please green readers, as will the images of London. This city has been thoroughly 'greened' (cf. the description of San Francisco in Callenbach's *Ecotopia*).

Twenty-first-century outer London is a mix of villages separated by blocks of woodland. From Chiswick to Putney there is thick forest. Hammersmith features 'sunny meadows and garden-like tillage': the Broadway is a mass of beautiful buildings rising up from the meadows. Hammersmith and Kensington are but two of the component London villages, set in the countryside and separated from each other by bands of woodland that run all over the old city.

And the nineteenth-century sprawl of houses built during Morris's day around Epping Forest, Walthamstow and Woodford, has been cleared (in 1955!). Beyond Aldgate the houses are dispersed in meadows, and the banks of the River Lea are again beautiful. East of the docks is flat pasture and a few houses set in the

> wide green sea of the Essex marshland . . . there is a place called Canning's Town, and further out, Silvertown, where the pleasant meadows are their pleasantest: doubtless they were once slums and wretched enough.

Central London is scarcely less idyllic. In Piccadilly, big houses stand in their own gardens; there are many fruit trees, orchards and tree-lined streets. Trafalgar Square, which has lost Nelson's Column and the rest of its concrete, is a big open orchard. While all the slums have been cleared from the inner city, some areas of dense housing are left in the business quarter; largely because they were so solidly built, and are roomy. The 'disadvantages' of dense living are here offset by splendid architecture – adornments and improvements having been added to the houses. The docklands are still in business, but not as intensively as in the nineteenth century. 'We have long ago dropped the pretension to be the market of the world . . .' and 'we discourage centralisation all we can'.

Where have all the people gone?

This question must nag at the mind of the socialist throughout such descriptions. There is more than a hint, in Morris, of the kind of elitism associated with the traditional romantic, who, while professing love of humankind, does not care to be surrounded by too many of them at any one time: derogatory references to the 'cockney' abound. We find some reassurance: the population of Britain in *Nowhere* is at the same level as the nineteenth century. 'We have spread, however', and helped to populate other countries 'where we were wanted and called for'! So, as with all golden ages, Morris's Britain is static, and although no birth control is discussed, the Malthusian potential for humans to increase their numbers geometrically – which seemed apparent in Victorian Britain – is not confronted. Neither, however, is any concept of a demographic transition, through universal affluence, discussed.

Kropotkin, by contrast, does take on, and repudiate, Malthus. In the tradition of eighteenth- and nineteenth-century philosophers of progress through science and technology, he thought that no limits to population growth were foreseeable, and densities of 600 people per square mile are quite possible. Through agricultural intensification, via a combination of technological advancement, labour-intensive cultivation and collectivisation, he believed that 200 families could be supported on 1000 acres. Britain could grow food for 90 million people, he argued (with an optimism which later on he came to moderate).

Agriculture, nature and beauty

The kind of collective farms which Kropotkin envisaged are mixed enterprises. His 1000-acre example is one-third in cereals and a little more in green crops and fodder, supporting thirty to forty milch cows and fruit (including two acres of glasshouses) and half-an-acre of flowers, with 140 acres set aside for public gardens, squares and 'manufactures'. The contrast with today's specialised farms could hardly be greater. Norfolk's 1000-acre ranches are often run by two or three people each. Their fields are empty except for huge machines. When you look at English farmland today, you see few livestock and fewer people, but both of these elements abound in Kropotkin's and Morris's rural landscapes.

Because of increased rotations, and the full use of farmyard and human manure, the contemporary problem of artificial nitrates, with its corollary of eutrophication of the waterways, would not apply. Perhaps this, and the loss of large-scale industry, is why the waters of Morris's Thames are clear and abundant in salmon.

The mental picture of Kropotkin's fields is less romantic than Morris's. It is one of intensive horticulture and market gardening, of the type which surrounded the nineteenth-century cities. The small fields yield highly, through high labour and sewage inputs, liberal irrigation, cheap glasshouses (today's polytunnels?) and heated soil. There are plenty of trees and hedges to protect plants and the soil, and many fruit trees and vines. Selectively-bred plants are sown widely-spaced to maximise yields. Kropotkin gives lengthy descriptions, drawing on extant French communes which used labour cooperatively and were surrounded by areas of densely-cultivated fruit and vegetable plots.

By contrast, Morris's farmscapes appear more relaxed and Constable-like. There are numerous references to haymaking, using people rather than machines, but beyond this what happens in the country is rather vague. It is looked after with great care to enhance its beauty and variety, and it is tidy. But this is not the tidiness of uniformity. So although, for example, willows are pollarded, it is not done to a uniform height in order to create that diversity which anarchists so value. In Morris's pollutionless world, there is, predictably, much wildlife: an increase in bird species, for example, including birds of prey. The banks of the Upper Thames are forested, wild and beautiful, having lost their 'gamekeeperish trimness'. People have a 'passionate love of the earth', and do not see nature as separate from themselves.

This enhanced sense of beauty is reinforced in human-made things. Human craft is seen in most objects – from tobacco-pipes to bridges and buildings. Gothic cast-iron bridges have been replaced by oak and stone ones. Big buildings are quaint and fanciful, with painted and guilded vanes and spirelets. Houses are low, and frequently of red brick and tiles, or of timber and plaster. And tumble-down ruins are not appreciated: 'we like everything trim and clean . . . like the medievals . . . it shows we have architectural power and won't stand any nonsense from nature in our dealings with her'.

Energy and transport

Neither author tells us much about the motive power for these quietly industrious societies. Morris simply informs us that power is available where people live, and it does not cause smoke. Windmills feature in Kropotkin's fields, to pump irrigation water. Morris's barges ply up and down the Thames with no visible means of propulsion. They are known as 'force vehicles'. For the rest, water transport is by rowing boat and sail, roads are still traversed by horse and carriage, and there are no railways.

Trade and international relations

Morris and Kropotkin agree on the vital anarchist principle that the nation state is an artificial device, whereby people are coerced into patriotism. Along with Marx, they see that the spread of capitalist commercialism undermines national and regional cultural variety, and want such variety to be re-established. In Morris's world, the system of rival and contending nations has simply disappeared, with the concomitant removal of inequality between people.

To Kropotkin, such a system is a nightmare, leading to war through battles for economic supremacy in a world market, and through the establishment of monopolies over trade, production and resources. But as each nation diversifies due to the spread of technology, and loses the advantages of commercial and manufacturing specialism, so self-sufficiency becomes essential and therefore large-scale international trade atrophies. Kropotkin accurately foresaw the 'de-industrialised' Britain which many would have us accept today as a fact of life, and, indeed, which liberal environmentalists welcome. However, Kropotkin does not follow exactly the Marxist line of analysis as to why de-industrialisation has happened. To him, it results from the 'inevitable' global spread of technological knowledge, aided by modern communications: to the Marxist it specifically relates to the capitalist firm's search for cheap non-unionised labour (Taiwan, Korea or Hong Kong, for example) and new markets, and is facilitated (rather than determined) by communications developments, particularly information technology. The Marxist would see increased global exploitation from an ever-powerful centre (Western-based multinationals) as the major result. Kropotkin, however, envisages that 'Industries of all kind will decentralise and are scattered all over the globe, and everywhere [is] an integrated variety of trades instead of specialism'. Each area therefore manufactures most of what it needs and makes *itself* its market; this in turn leads to rising affluence, and greater material uniformity. It may be deduced from this that regional and national differences in landscape consequent on core–periphery economic contrasts will be a thing of the past.

The kind of polarisation that we witness today, between Britain's or Italy's north and south, or North America and 'black' Africa, would disappear, along with the concept of landscapes of affluence and landscapes of material want and spiritual despair. Certainly no such regional differences are apparent in the visions of Kropotkin or Morris.

4.5 THE FOURTH WORLD

Principles of bioregionalism

The connection between anarchism and the green movement now extends beyond utopian and fanciful thinking, into the (urban and rural) communes movement: people who are 'doing it'. The fourth world movement encompasses activists living in communes or neighbourhood groups (particularly in the USA) and all others who, via theory or practice, push the politics of regional and local separatism.

The 'fourth world' is defined in Sale (1985, 156):

Table 4.3 Bioregional compared with industrio-scientific paradigms
(after Sale 1985, 50)

	BIOREGIONAL	INDUSTRIO-SCIENTIFIC
Scale	region	state
	community	nation/world
Economy	conservation	exploitation
	stability	change/progress
	self-sufficiency	world economy
	cooperation	competition
Polity	decentralisation	centralisation
	complementarity	hierarchy
	diversity	uniformity
Society	symbiosis	polarisation
	evolution	growth/violence
	division	monoculture

just as exterior colonies broke away from empires to form the third world, so internal colonies – the fourth world – are now trying to break away from states.

(This is quite different from some other definitions: Bullock and Stallybrass (1977), for example, describe the fourth world as third world countries which are not rich in natural resources like oil.)

The fourth world movement seems to span the deep (Gaian) ecology – social (communitarian) ecology 'divide'. It, like the wider green movement, has many apparent similarities with anarcho-communism, especially in the details of the world it pursues, but there also are some important political and philosophical differences between them. 'Bioregionalism', on which the movement is based, often veers towards liberal and libertarian anarchism.

Sale, who identifies sixty bioregional groups in 1985 in North America, lays down four principles of bioregionalism. Two link firmly with anarcho-communism. These are, first, *liberating the self*: reducing the importance of impersonal market forces and bureaucracies, opening up local political and economic opportunities, enjoying communitarian values, of cooperation, participation, reciprocity and confraternity, and having roots. Second, there is *developing the potential of a region towards self-reliance*. The other two lean more heavily on nature philosophy, a theme which we have already identified as generally muted in anarchism, but which revives echoes of Thoreau. *Knowing the land*, involves walking it, to 'become conscious of the birdsongs and waterfalls and animal droppings' and studying 'optimal settlement areas' to identify regional carrying capacity. Sale admits that this is a 'bit bucolic'. *Learning the lore* includes knowing folklore and history, the many useful technologies that traditional people had, and the 'traditional wisdom of mankind'.

The 'bioregional paradigm' thus differs hugely from the present 'industrio-scientific paradigm' (see Table 4.3), elevating 'natural systems' to become a 'source of nutrition and of metaphors to sustain our spirits' (p. 21) and emphasising sense of place, ecological consciousness and a feeling for 'bioregional spirit', as befits deep ecology. It revives old 'regions' such as the 'Shasta nation' (south Oregon and north California – Callenbach's Ecotopia), or 'Middle England', a pre-Norman–English concept. The bioregion is 'Any part of the earth's surface whose rough boundaries are determined by natural characteristics rather than human dictates' (Sale, p. 55). Bioregions are found in size hierarchies: an 'ecoregion', such as the Ozark Plateau, is several hundred thousand square miles, and is defined by 'native' vegetation and soils. A 'georegion', such as California's Central Valley, is defined especially by its physiography; for instance river basins, mountain ranges and watersheds. Georegions have 'distinctive boundaries'.

'Obviously this business of bioregional identification is no simple matter, but the broad definitions are clear enough to anyone who will look', says Sale (p. 59), apparently discounting the difficulties of generations of geographers who have tried to identify borders of such 'natural' regions. They have largely abandoned the task because of its questionable usefulness, and the difficulty of being precise. (Sale thinks there is an advantage in keeping borders vague: it would blunt possessiveness and defensiveness.) And they have found the whole concept of 'natural' regions increasingly untenable as it largely overlooks the reality of an urban oriented (if not based) population, and of increasingly universal economic and social processes which homogenise places.

> The regionalists' error . . . was the notion of the uniqueness of location. But there was another dimension to the critique . . . the presumed intimacy of these ecological bonds was admirably suited to the historical geography of Europe before the industrial revolution . . . with the final disappearance of the old, local, rural, largely self-sufficient way of life the centrality of regional work to geography has been permanently affected.
>
> (Gregory, in Johnston 1981, 287, citing other geographers)

Where there still is regional differentiation in industrial society it is based far less on watersheds than on economic function, and other criteria such as language and religion (see Alexander 1990). To suggest re-identifying and reinstating old cultural, let alone physical, regions appears grossly unrealistic. The fantasy in bioregionalism is brought home in Devall and Sessions's view (1985, 24) that the 'sense of place' which can be 'recultivated' in the city depends on identifying geology, plants, animals and landforms 'buried under the mass of concrete that forms the modern image of the city'!

Undeterred by such objections, bioregionalists suggest many potential advantages of their schemes; ecological, economic and social, all of which appear to be regarded as automatic outcomes of the act of changing spatial forms and scales of organisation. The disadvantages of the city, as an allegedly

inefficient energy user and waste recycler, will be overcome, says Sale (p. 65) –
to whom there is 'no doubt' that a city of 500,000 or more 'has gone beyond the
ecological balance point at which it is able to sustain itself on its own
resources'. By contrast (p. 76), the small communities in self-sufficient
bioregions would be insulated from boom-and-bust economic cycles initiated
from far away, being free from distant economic control. They would be richer,
not having to pay for imports or high levels of transport. They would control
their own currencies and economic policies (cf. the local 'green' currencies and
unofficial employment and training exchanges currently being tried – see
Ekins 1986, or Wilding 1991). Their people would be healthier, more
'cohesive' and self-regarding. Their locally-made decisions would be based on a
cooperation born of sharing the same problems, and would stand 'at least a fair
chance of being correct' (p. 95).

This *areal division of power* is what will best promote equality, efficiency,
welfare and security in all society, say bioregionalists. More cohesion and
alliance mean less crime, more citizen participation in government and
sensitivity to the needs of others. At the same time, regional cooperation will
inevitably be encouraged by the needs of water and waste management, and
inter-regional transport and food distribution.

For bioregionalists, merely revising the scale of living will solve, at root,
many 'abstract and theoretical problems' because 'in small communities people
will see *the effects of their own actions*, e.g. on the environment, and therefore
the practical wisdom in mending their ways. It is thus not a question of change
by changing ideas or morality first. Furthermore, as Papworth (1990) suggests,
it will militate against destruction and for the development of human
creativeness: 'The only form [of power] which holds aggression in check and
liberates the creative genius of the human spirit is not united power but
equally divided power'.

The bioregional and anarchist society

When more details emerge of how bioregions would look and be organised,
they accord closely with anarcho-communist visions. Sale's (1985, 85–6) list of
the features of bioregional production, though he makes no mention of
Kropotkin and anarchism *per se*, would substantially appeal to a Kropotkinite,
or indeed an SPGB member. There will be production for need, value according
to social usefulness, labour without wages, common ownership of the means of
production and a planned economy, to produce enough for everyone. The
wealth of nature would be the wealth of all, through common ownership of the
commons. Though each bioregion would trade as little as possible, relying on
natural assets and finding substitutes for absent materials, there would be
federation into 'morphoregions' to sustain hospitals, universities and sym-
phony orchestras. Similarly, in its publicity leaflets, the anarchistic *Movement
for Middle England* wants to encourage local moots of around 50 households as

Table 4.4 Anarchistic features of alternative communities identified by Rigby (1974)

The main dynamic of survival is cooperation/altruism
The group rather than the individual is the social unit
They are not run by any authoritarian body
Cooperation is voluntary, not forced
Individual rights and freedoms are respected – perhaps becoming almost sacrosanct
There is no property or money system – or a reduced system
Free distribution of commune products
Life is economically/socially simple – non-materialistic
Behaviour is satisfying rather than optimising
Flexible work/leisure system away from capitalist work routines
No or few drugs
Healthy lifestyle – responsibility for health with the individual, not the welfare state
Technology is related to *need* and the community (i.e. soft energy/organic garden-
    ing–vegetarianism/group transport)
Minimising alienation (related to preventive health care)
Creative work, self-made amusements
Full participation in running the commune: decisions by consensus, not voting

the basic building block of future democracy; delegating upwards to neighbourhoods, districts, counties/areas, and regions in due course'.

Chodorkoff's (1990) social ecology model of community development also has similar features, drawing openly on Owen, Fourier and Kropotkin. It recreates 'pre-bureaucratic face-to-face ties' in town meetings, block associa-tions, neighbourhood planning assemblies and popular referenda. Its economy consists of food and producer coops, land trusts, credit unions and common lands. It seeks to rebalance country and city, in the apparent (mythical?) belief that they respectively constitute *gemeinschaft* and *gesellschaft* societies. And it fosters senses of scale and place, 'active' relations with nature and an organic relation with environment.

The 'bioregional mosaic', then, 'would seem to be made up basically of communities, as textured, developed and complex as we would imagine' (Sale 1985, 66). This is the basis of social ecology too, and of the more radical interpretations of the sustainable development paradigm: 'Earth is best described as a mosaic of coevolving, self-governing communities' (Engel and Engel 1990, 15). This appeal to community, in the specific form of neighbourhood unit, block or street, or urban or rural commune, is one of the uppermost themes common to anarcho-communist and green writing (see for instance Bahro (1986), Goldsmith (1988), Robertson (1983), van der Weyer (1986), and Schwarz and Schwarz (1987). Indeed, countercultural communes, especially those founded in the 1960s and early 1970s, seemed in many ways to be anarchistic (see Table 4.4). While some have now lost part of their anarchism and radicalism in the 1980s and 1990s, they have developed ecocentrism, while a third of communards still have clearly enunciated anarchist–socialist beliefs (Pepper 1991). Not all anarchists, however, are

189

committed to the commune. Purchase (1990, 10), who rejects anarchism based on 're-tribalisation of society' believes that

> Although anarchists would certainly like to see small villages and towns become independent, a return to small scale and essentially isolated communal lifestyle on a mass scale is both repugnant and implausible. Anarchism is not a backward looking pre-industrial ideology.

The natural world as a model

Chapter 4.2 explained how aspects of the appeal to a 'natural', 'organic' order, where people must model their society on 'nature' have distinctly reactionary implications, from which Bookchin and anarcho-communists like Purchase are to be distanced. But much of this ideology is strongly present in bioregionalism, through its deep ecology aspects. Indeed, Devall and Sessions (1985, 21) clearly describe the bioregion as a conservative concept, which has been the 'animating cultural principle through 99 per cent of human history'.

That the 'natural world' and 'traditional/natural' societies form a modern social model to Sale is clear. For him, the very facts that tribal/preliterate societies – 'most people since Cro-Magnon man' – have always lived in small groups, and have been self-sufficient seem adequate to justify our doing the same today. And since at ecosystem level self-sufficiency is possible because there is enough variety of species and enough territory, 'The same rule seems appropriate for human self-sufficiency as well'. The phrase 'seems appropriate' is used in lieu of any more argued justification. Sale embraces Goldsmith's brand of ecological energetics, modelling society on 'laws of ecodynamics' which state that conservation is the basic goal of the behaviour of structures and matter, and that natural systems tend to stability and climax: hence bioregional economies must conserve resources, relations and systems of the natural world and have stable means of production and exchange. And in bioregions there must be 'cooperation', by which Goldsmith means people occupying happily a 'niche' allotted to them, and fulfilling their functions. These appeals to equilibrium and stability are deeply conservative because they would freeze society in given state, or at best favour only slow 'organic' change. Thus the prospects of eliminating that society's existing imperfections would lie remotely far away. By contrast, perpetual change and progress, the cornerstones of modernism, may have all sorts of undesirable side effects. And there are many who argue, with ecocentrics, that they constitute an inherently harmful development model. However, this model does always hold out the possibility of social improvement within foreseeable time, hence it can be the only one which appeals to would-be social revolutionaries.

Sale's bioregionalism fundamentally hinges on reviving the 'Gaea' concept, which, he says, stretches back 25,000 years and embraces vitalism, nature worship, animism and the natural community. We need to study diligently nature's laws and (p. 55) the first law of Gaea is that the face of the earth is ordered '. . . not into artificial states but natural regions . . . the natural region is the *bioregion* . . . defined by the givens of nature'. There is (p. 59) an obvious second law of Gaea – all life is divided into biocommunities, and the fact that Indian tribes distributed themselves in bioregions demonstrates how 'well grounded' bioregionalism is.

This echoes Leopold Kohr's (1957) views, which inspire much of the fourth world movement. Kohr says (pp. 97–8) that breaking nations into smaller units is

> not only a matter of expediency but of divine plan, and . . . it is on this account that it makes everything soluble. It constitutes, in fact, nothing but the political application of the most basic and organising balance in nature. The deeper we penetrate into its mystery the more are we able to understand why the primary cause of historic change . . . lies not in the mode of production, the will of leaders, or human disposition, but the size of the society within which we live.

Thus, besides embracing a rather old-fashioned concept of regional geography, openly picking up on the traditions set by Vidal de la Blache, Ratzel and Le Play, bioregionalism resuscitates the somewhat discredited theme of environmental determinism, which infused much geography up to the 1960s. Morphoregions, says Sale (1985, 58), are

> identifiable by distinctive life forms on their surface: towns and cities, mines and factories, fields and farms, and the special landforms *that give rise to these particular features in the first place* [emphasis added].

Bioregions (p. 55) are 'distinguishable by particular attributes of flora, fauna, water, climate, soils and landforms' and 'by the human settlements and cultures those attributes *have given rise to*' (emphasis added).

Just how politically reactionary Sale's bioregionalism really is, and therefore divergent from anarcho-communism (and Bookchin's own social ecology), is an open question. While on one hand it echoes much of Goldsmith's general appeal to conservatism, tradition and biological reductionism, on the other it repudiates any idea of hierarchy such as sometimes appears in Goldsmith's notions of niche (i.e. the chain of being). Sale insists that niche and mutuality in nature mean 'heptarchy' – distinction *without* rank. Similarly, in traditional societies important people did/do not have *personal* power: Apaches did not have 'chiefs': it was the white settlers who thought that if these were military leaders then they must be 'chiefs'.

Yet there is an 'apoliticism' running through bioregionalism which, by its lack of appeal to radical change, or its naivety, is potentially reactionary. Thus

> 'Sacred place' and 'earth-bonding' rituals are less the inventions of displaced natives than the romantic contrivances of well-meaning

outsiders. For the native peoples of the South-West [USA], the key to bioregional renewal lies not in some romantic return to 'sense of place', but rather in the struggle to regain access and control over land and water rights that were expropriated by capital and the state through violence and political/legal chicanery.

(Pena 1992, 89)

But this would not constitute the 'slow change' which is the key in bioregionalism (Sale 1985, 120) and which says that excessively rapid change begets crime and violence. Kohr's purpose is to eliminate violence and war – by eliminating the great powers themselves. However, he does not propose to replace any other possible causes of war: no form of government, he believes, has inherent superiority over others, so there should be freedom of choice as to whether to have communism, dictatorship, monarchy or republic (p. 109).

And judgements about morality are also lacking in Sale's work. Morality, it seems, is simultaneously optional, natural and heavily punished if absent. It is not a question of 'right morality' but of understanding one's own part in causing environmental problems which will make one do right – both as a matter of practicality, and of spiritual sensitivity:

When [people] look with Gaean eyes and feel a Gaean consciousness, as they can do at the bioregional scale, there is no longer any need to worry about the abstruse effluvia of 'ethical responses' to the world around.

(Sale 1985, 54)

Yet at the same time 'Of course the entire moral structure of an ecologically conscious society would rest on Gaean principles' (p. 120), therefore ignorance of the phrases 'carrying capacity' and 'biotic community' would be a crime and most punishment would go to those who damage the ecosystem. Hurting *people* is apparently different however, for, because of the diversity and self-determination principles, not every bioregion would be 'likely to heed the values of democracy, equality, liberty, freedom, justice and the like'. Still, the system will work even if not all in it are good: it has structures which will 'minimise aberrant behaviour'.

If this smacks of the language of ecofascism, so do other aspects of the nature-as-template approach. Thus, war is avoidable in a small-scale organisation, hence any tendency to grow large must be offset through the (biological) process of dividing off and creating new communities through 'voluntary resettlement'. This 'as the experience in several European countries has shown, is not a difficult social policy to carry out'. In fact such experience suggests that it *is* difficult, since it frequently is not voluntary, and/or there is resistance from the host region.

Perhaps there is just plain naivety here, as there is in the notion that bioregionalism can unite 'The National Rifle Association Hunter in Pennsylvania with the environmentalist in Colorado, both of whom understand the balances of nature' (p. 168). Or perhaps it is that *libertarianism* rather than anarchist–communism is what bioregional sociology is about, coupled with an ecological moral authoritarianism. 'Libertarianism', according to Sale (p. 90) is the 'closest word in our inadequate vocabulary' to describe bioregionalism's polity. It means:

> An extreme version of liberalism, hostile to all forms of social and legal discrimination between human beings and favouring the absolute minimal constraint by society on individual freedom of action.
>
> (Bullock and Stallybrass 1977)

Yet this still sits ill with the bioregional land ethic – that involves heavy moral obligation to all organisms and their interactions, that is 'the ecosystemic order itself' (Engel and Engel 1990).

And there is the added danger that this ethic will amount to little else but a different form of nationalistic sentiment. For it is based on familiarity and affection for the background of one's youth, reinforced by myths about one's region from education and socialisation. If, as such, it breeds oppressive attitudes towards other regions, it is no more acceptable to the anarchist–communist than national jingoism, even though it does originate in common, community feeling and is independent of the state (Cullen 1991).

That said, a review of bioregional magazines such as *Fourth World Review* (edited by John Papworth) or *The Regionalist* (produced by various people) shows that they both steer clear of regional chauvinism, seeing themselves perhaps as more of a defensive movement against centralised aggression, and making a virtue of the paradox of lacking a coherent set of social ethics, yet having a quite specific view of the social and spatial forms that would follow from an absence of state-centralised organisation.

Happening now?

Another element common to anarchism and fourth world ideology is the assertion that their preferred society is already emerging from the present one, and that signs of this are all around. This stems from their shared belief that anarchism or bioregionalism are *natural* tendencies – part of the 'natural' society – which therefore inevitably keeps re-emerging in all but the most adverse circumstances (despite the huge amount of 'unnatural' behaviour). Kropotkin's *Fields, Factories and Workshops Tomorrow* reiterated this theme. Kropotkin believed in 1899 that decentralisation of population and industry were detectable in Britain and world-wide, and that most production was already in small units, e.g. 'nearly one-fourth of all the industrial workers of this country are working in workshops having less than eight or ten workers per establishment' (1985 edition). And in the 1974 edition, Colin Ward added substantial notes and commentaries to support the assertion that dispersal of production and people were happening then.

Similarly, Sale (p. 151) believes that bioregionalism is 'in sinew and blood thoroughly expressive of the basic trends of the late twentieth century', i.e. deep environmental concern, distrust of bigness and determination to form a people's own black economy in response to the disintegration of the industrial economy and the nation state. Chodorkoff (1990), meanwhile, provides many examples of how urban residents' groups in American cities are helping themselves, without state aid, through creating green spaces, adventure playgrounds, allotments and so forth, from waste ground.

What is happening in late-1980s' and early-1990s' Europe, east and west, appears to be that movement to re-establish the old small nations which Kohr advocated in 1941. Asturia, Castilia, Catalonia, Croatia, Bosnia and Serbia, for instance, have reasserted themselves. Papworth (1990) is rightly gloomy about Western European bioregional prospects, in the face of an inexorable counter-trend towards European integration. Yet he euphorically cites the break-up of Eastern 'communism' as an expression of natural bioregionalism.

Unfortunately, it is highly doubtful that this, or any other of these things truly represent anarchism or bioregionalism. Decentralised, federated Eastern Europe and the USSR seem at present hell-bent in exchanging their state-centralised yoke for another equally burdensome one in the long run; that of private capital, which has repeatedly demonstrated for the past three hundred years that its cherished trickle-down theory of wealth (and power) does not work for the vast majority of the world's people. Industry may be flung all over the world, but the strings which control it have a longer and longer reach, and are pulled by fewer, not more, hands, from the headquarters of European and American multinationals. People may be physically forsaking the city in the West, but they still depend on it. Capitalism may be alienating people so that they may form black economies, but it still holds the reins of political, coercive power firm.

4.6 ANARCHIST APPROACHES TO SOCIAL CHANGE

Social change theory and the state

Anarchism attributes social and environmental ills primarily to the growth of relationships of hierarchy and domination among people. Quite *why* this is thought to have happened is often obscure. 'Human nature' is not generally blamed; indeed the 'natural' society is thought to be based on mutuality, cooperation and freedom for all – a natural state, then, from which many have temporarily deviated, but which is reasserting itself. Neither are relations of production, geared to specific modes of production and the class systems they create, blamed *at root*.

Fundamental causes are deemed to go beyond class, specific cultures, or economic systems – they are more universal. Some anarchist–feminists, for instance, see domination and oppression as inherent in relationships between the sexes, but this does not take us much further in the search for the roots of such 'unnatural' behaviour.

In this way, the anarchist view of why things are unsatisfactory, and how they may be changed, tends to be ahistorical. So much do anarchists reject historical determinism, it seems, they have not developed any theory of history at all.

But this is not quite accurate: most anarchists have a strongly evolutionary historical perspective. It underlay Kropotkin's work, reappeared with Mumford and surfaces strongly in Bookchin's social ecology. As Clark (1990a) describes it, evolution is a process of increasing diversification through history, where ultimate dynamic equilibrium and harmony (homeostasis) will eventually be achieved through 'ever changing differentiation' towards increased 'value' and self-organisation. This view echoes nineteenth-century biological and social (cf. Herbert Spencer) evolutionary biology. But it rejects historical determinism; describing merely a 'tendency' towards these things but not an inevitability.

An evolutionary perspective strongly incorporates, in Bookchin's (1990) dialectical view, ever increasing freedom for humans, in the way of *self*-determination. This is partly predicated on cultural and technological, as well as political, development. Following Mumford (and Marx) Bookchin does not see all such advance as a distancing us from nature and 'naturalness'; rather it is the natural process whereby second evolves from first nature (see Chapter 4.2). Taking a side-swipe at deep ecologists, Bookchin (p. 210) says:

> A humanity that fails to see that it is the embodiment of nature rendered self-conscious and self-reflexive has separated itself from nature, morally as well as intellectually.

Nonetheless, conscious that anarchism seems ahistorical to many, Carter (1989) attempts to devise an anarchist theory of history to rival that of Marx. He asks why Marx's 'three tiered' model cannot act in the reverse direction. Instead of the mode and forces of production at the economic base explaining the relations of production, which then explain superstructural phenomena (non-economic institutions, including the state), might not, conversely, the superstructure independently *select* specific relations of production, and therefore the mode and forces of production? Carter thinks that it might, meaning essentially that we suffer the existing mode, forces and relations of production directly because of the machinations of the anarchist's *bete-noir*, the state. He reasons deterministically, and rejects any more sophisticated 'dialectical' view of this chain of cause and effect, when, as we have noted, he brands the whole concept of 'dialectics' as sophistry: a mystification of a simple systems view.

His alternative, anarchistic, model (Carter 1988) in which the 'politico-ideological substructure' of fundamental beliefs has primacy over economics, does not, then, regard the state as a mere instrument of the dominant economic class, which will wither away when a classless society – predicated on a socialist mode of production – is attained. Rather, it has a life of its own, its own interests, and its own actors, who will frequently, but coincidentally rather than necessarily, side with the bourgeoisie when to do so will result in their own aggrandisement. However, they are also sometimes disposed to side with the proletariat *against* the bourgeoisie should this be more in their interests. They are less likely to identify with a specific economic class, then, than with a set of ideologies which they favour (such as the belief in property rights).

For the state wants whatever will foster the maximum development of productive forces, because this maximises the production of a surplus, which can then be used to build the forces of coercion (military, police, etc.), which the state uses to sustain and protect itself. The implication for anarchist strategy is that a social change to egalitarianism cannot be effected by taking over the state and using it, for it will always act in its *own* interests. And furthermore it will take counter-revolutionary action against the anarchists' attempts to create egalitarian relations of production, because egalitarianism would yield no surplus which the state could then appropriate and use. Hence revolutionary change must be founded on demolishing the state *from the outset* and the ideologies on which the state, and the economic system, are predicated. 'It is highly implausible that a non-participatory state would endorse the introduction of significantly participatory or egalitarian relations of control', Carter (1991) argues, when developing this anarchist theory also into a green political theory.

This theory is essentially a sophistication of an old anarchist theme, which is the 'principle of the unity of ends and means without compromise' (Sekelj 1990, 20). As Neville (1990, 6) expresses it

> One does not obtain freedom by replacing tyranny such as the tyranny of the Tsar with the so-called 'dictatorship of the proletarian' . . . nothing is changed in this circumstance except the label on the bottle. True social revolution, according to Proudhon, must never be created on the foundation of hierarchy and leadership. Only a decentralised society – both politically and economically, without the need for leaders, can be classless.

In short, to create a free, unrepressive, peaceful, classless society, one must use free, unrepressive, peaceful and classless *methods*.

This view originates from Bakunin's anarcho-communist model of revolution. It consists of a short stage, of pulling down the existing order, and a longer one, where – as an *act of revolution* – the kind of society which is ultimately wanted is itself set up. There is no 'transitional' stage, containing features which would not be in the ultimate society. 'The significant difference of anarchism', says Sekelj (p. 20), 'is that even in revolution itself all forms of the principle of government are abolished, and the federation of communes and voluntary productive associations established'.

Thus, as part of the process of getting there, you create for yourself what it is you want to attain, within the society that you wish to replace. This is the idea of 'prefiguring' and it is the basic anarchist theory of social change; *via action*. The differences within anarchism mainly revolve around (a) how much this is achieved by attempting Bakunin's first stage of destroying the existing order, or, by contrast, simply ignoring and 'bypassing' that order; (b) to what extent individualist or collectivist strategies are followed; (c) how much to base the latter, particularly, on 'modernist', class struggle and class analyses, or on 'postmodern' approaches, through new social movements or through rejecting 'the scene' altogether (i.e. destroying the game by refusing entirely to play in it).

Getting there: trades unions and anarcho-syndicalism

'Social anarchism, in both theory and practice, has always acknowledged the necessity of working class or trade union organisation' says Purchase (1990, 12). For one thing, unions can

> boycott the use or reckless disposal of dangerous substances and implement newer and safer techniques and procedures without waiting for the government to pass new laws on industrial regulation. . . . At such times the people begin to organise themselves without paying heed to what the government may or may not say. The fall of government and the nation state becomes inevitable.
>
> (ibid., 8–9)

Here, Purchase clearly envisages a destructive stage of revolution, in which workers are involved as part of class struggle. In addition, he argues (p. 13), they, 'the producers of all social wealth' are 'ideal vehicles for a host of economically vital inter-communal activities' in a prior prefiguring stage. For

> ordinary working people must develop non-bureaucratic and directly democratic forms of agro-industrial organisation, *in advance of the revolutionary movement*, capable of ensuring that vital services function efficiently . . . from the moment the state-capitalist order begins to disintegrate. Without agro-industrial working class or trade union organisation, revolutionary anarchism will remain an intellectual fantasy and a philosophical pipe dream.

Purchase's perspective draws on contemporary Australian experiences of anarcho-syndicalism and worker action to secure anti-development, pro-environmental objectives (Chapter 5.4). In the US and Europe, however, anarcho-syndicalism is largely an approach of the past. It dominated the French and Spanish labour movements from 1902–14, and was a mass movement aiming to abolish government via a general strike and to manage society by the working class. The syndicalist Confederation Général du Travail

Table 4.5 Russell's anarcho-syndicalist (guild socialist) vision of the world 'as it could be made', 1918, chapter 8

Neither capitalism nor state socialism
Communal land ownership
No-one compelled to work: a basic income to all
Necessaries freely distributed, luxuries purchasable with money
Four-hour working day for those in work
All domestic work is paid work, comparable to industry
Every industry is internally self-governing by workers' councils
Management is elected by workers
Relationships between producers are settled by a guild congress
Compulsory education to age 16: free education up to 21 years
Everyone is taught several trades or skills
Education is to follow children's instincts, and not to instill into them reinforcement
    of the socio-political status quo
Marriage is free and spontaneous: not sanctified by priests or formalised by law
Government and law are minimal: property laws are obsolete, murder is forbidden
There is a state, but most power is devolved to communities
There is disarmament, and a league of nations
No nation has its own army: each contributes to the league of nations.

---

cut all links with political parties in favour of economic direct action, including work-to-rule, strikes, machine breaking, consumer boycotts, intimidating strike breakers and violence against the bosses and their wealth (Harper 1987, 68).

The movement influenced American and British labour – the latter devised a diluted form of anarcho-syndicalism known as guild socialism, which Russell saw as the ideal anarchist/socialist compromise (see Table 4.5). Marxist socialism was, he felt, ultimately bourgeois in origin and outlook, anarchism was aristocratic. Only syndicalism was indubitably labourist (1918, 67).

He explained the basis of syndicalism as collective ownership of the means of production through organised labour (p. 66). It stood for industrial as opposed to craft unionism. The unions would organise production in each workplace, federating twice over: with all other 'syndicats' in the same locality, and with all similar trades elsewhere. Here, then, is a form of small-scale and local organisation which is economically focused, and centrally concerned with production. By comparison, American–European green anarchism today often emphasises the rural commune, the community as a place of living rather than, foremost, of work, and the individual or group political power as consumers rather than producers. Perhaps an infusion of Australian-style anarcho-syndicalism into the American–European green movement might shake it up today, just as, apparently, syndicalism shook up the labour movement in the 1910s. It did this, said Russell, because of its emphasis on 'man as *producer rather than consumer*', its revival of the quest for liberty – it was more concerned for freedom in work than material wellbeing – and its demand for fundamental social reconstruction.

Russell's guild socialism envisaged that the state would continue, but in a relatively minor role. It would own the means of production as a trustee for the community, and through Parliament it would represent the community-as-consumers. The community-as-producers would be represented by a National Guild: a federation of all the local factory guilds. These two bodies would be overseen by a joint committee of producers and consumers, to fix taxes and prices, and administer what little law and order forces there would be. This somewhat bureaucratic 'anarchism' is less interesting for its detail than for the fact that it raises the issue – central to anarchists and socialists – of whether centralised coordination and organisation needs, *per se*, to be bureaucratic, faceless, hierarchical and oppressive. Clearly, in this model, in the form of the National Guild, it does not, but then how does the parliament sitting alongside it in this properly 'democratised state' differ from this guild? One can hardly regard 'the state' as the invariable repository of all that anarchists revile, yet posit its existence *alongside* similar but acceptable anarchist federal bodies.

In this way, the guild socialist model indirectly raises the possibility that 'the state' does not *have* to be oppressive: benign forms are envisageable. (Paradoxically, the socialism of the SPGB, however, rejects this suggestion on the grounds that the state is *by definition* a force for coercion.) By the same token, Russell's model leads one to reflect that federal bodies such as the National Guild might not be automatically exempt from all the ills associated with giantism, remoteness and bureaucracy, which anarchists resolutely lay at the door of 'the state'.

Getting there: communal and lifestyle anarchism

Anarcho-syndicalism is a socialistic anarchism, emphasising collectivity, the material economic base, and trades unions. By contrast, communal and lifestyle anarchism, which is what much green anarchism amounts to today, has a more liberal ideology. This emphasises the role of the *individual*: even in anarchistic green communes. Social change is seen to start from the attitudes and lifestyle of the individual, whose development and self-knowledge is sacrosanct (Pepper 1991, Rigby 1974, Abrams and McCulloch 1976). Also, social change – revolutionary or more gradual – is mainly thought to proceed through bypassing the state, not attempting directly to overthrow it first. Communes, coops, squats, local currencies and the like are all seen as prefigurative green anarchism, undermining capitalism by setting an example of sound and preferable lifestyles, which others will see and want to follow. This perspective tends to dismiss class struggle, the labour movement and conventional politics, and, in practice, finds it difficult to subsume the individual self to a collective self. While it searches, idealistically, for a *gemeinschaft* solution, its adherents come from a *gesellschaft* society and are steeped at root in the mores associated with it, even though they wish to renounce them. As Dobson (1990, 145–6) expresses it:

199

A general problem with the strategy of lifestyle change is that it is ultimately divorced from where it wants to go, in that it is not obvious how the *individualism* on which it is based will convert into the *communitarianism* that is central to most descriptions of the sustainable society. . . .

Here is an ambiguity which this brand of anarchism shares with much of the green and feminist movements (see Hallam and Pepper 1990). It can become so skewed towards the idea of the reform of the individual's values and lifestyle as the primary political route to radical social change that it ends up seeming positively antipathetic to the notion of the collective. For instance, Roszak's (1979) anarchist–ecological view says: 'Persons come first before all collective fictions'. There is no shelter in revolutionary mass movements, and the counterculture's concern for the individual is 'beyond class struggle'.

Communal green anarchism seems to owe much to what Joll (1979) describes as that desire for simplicity and frugality, without the technological achievements of the machine age, which many anarchists share (though not, presumably, the likes of Mumford, Kropotkin or Bookchin). It tends towards romantic idealisation of the rural and the non-industrial, though its adherents are often aware of the dangers of such thinking and steer clear. However, the decentralised community, with its simple rural life and small-is-beautiful philosophy is, says Joll, a common outcome of the prefiguring approach. It foreshadows many of the elements of that society it seeks to replace: pacifism, non-sexism, non-hierarchy, sharing, spontaneity, ecological harmony, the extended family and the local ('black') economy. At least, such is the intention if not the reality (see Rigby 1990).

Not all of these anarchists oppose the idea of 'direct action' against the state. Some few carry out symbolically important and spectacular acts, such as damaging nuclear missile-carrying aircraft, invading military bases, or 'defacing' advertisement hoardings with graffiti which reveal the sexist nature of their content. However, rather more interpret 'direct action' in a more anodyne way – following, for instance, Devall and Sessions's (1985) deep ecology injunction to write poems, do yoga and breathing, live simple lifestyles, and indulge in earth-bonding rituals – all as forms of 'personal direct action'. The *Green Anarchist* magazine itself, uncompromisingly enjoins its readers to

> deny the state its tax income by building an alternative green and black economy through tax evasion and buying untaxed goods, allotments, guerilla farming and squatting and alternative medicine, education and energy to sustain our needs. . . . We must live free of exploitation, boycotting banks and multinationals.

But then, more lamely, it goes on to declare that 'Personal change is not enough. We must welcome all wanting liberty and unite as a culture of resistance through festivals and magazines' (1991, no. 28, p. 2).

As Ashton (1985, 18) points out: 'the idea of moral transformation to strengthen the individual's ability to transcend the status quo' is shared by greens and many on the left today

> who subscribe to ideas associated with the slogan 'the personal is political'. It has a long history. . . . From Christianity to Gandhism, Owenite Socialism to contemporary communes and GLC economic initiatives the 'exemplary project' and moral stand have been deployed to effect social change. This form of practice is not to be devalued . . . [though it has won] frequently only short-lived victories.

Getting there: new social movements and coops

The rejection of conventional collectivist strategies by many anarchist greens today reflects the aversion to Marxist class struggle noted earlier. Bookchin is more averse than most, being influenced by the Frankfurt School (discussed in Chapter 3). Like many of its adherents, he rejects the working class as social change agents, in favour of new social movements, particularly greens and feminists, as the real repositories of radicalism today. They comprise, he thinks (1989, 271), many thousands of anti-capitalist, conspicuously decentralist people, committed to direct action. They are neo-populist, more anarchist than socialist and more akin to

> Kropotkin's muscular notions of a radically decentralised society and its complete rejection of capitalism than Schumacher's Buddhist economics, which includes vague compromises with a market-oriented society.

Unfortunately, Bookchin's analysis here seems to lack the more studied approach to new social movements of, say, Alan Scott (see Chapter 3). For as a description of the green movement in general it seems very inaccurate: even Hunt's radical green anarchism, described earlier, includes compromises with a market society. Bookchin heaps praise on the German Greens for their brilliance and radicalism, but following their near elimination from the national parliament in 1990, there is a 'catalogue of principles compromised and ideals ditched' which they stand accused of by social ecologists (Biehl 1991). Contemporaneously, the British greens appear to be sliding towards the 'realist' and 'managerialist' camp, aping all the main political parties (Wall 1991).

For Cahill (1989, 232), a more promising manifestation of the new social movements is the

> recent upsurge of coops . . . which are, themselves, anarchism in its latest political manifestation. . . . Contemporary coops, and the support

structure which has grown up around them are subtly imbued with the anarchist spirit.

(ibid., 251)

Indeed, they have five elements which are also found in anarchist economic organisation: decentralisation, egalitarianism, self-management and -empowerment, based on local needs, and supported non-hierarchically by other autonomous units. Cahill sees them as ways of protesting against capitalism and dealing with the effects of de-industrialisation. But, more diffidently, he acknowledges that they are also favoured by pro-capitalist political parties, and have been a 'third sector of capitalist economics' from which they now 'need rescuing'.

Indeed it is difficult to be convinced that coops are truly prefigurations of an anarchist society, rather than yet another self-exploitative device through which capital shuffles off its social and community responsibilities (part of the Thatcherite 'stand-on-your-own-two-feet' philosophy). Cahill's evidence is tenuous:

What is there to suggest that coops as they exist now are part of an economic alternative? My strongest evidence is a hunch, an intuition shared by most of the cooperators and cooperative developers to whom I have spoken.

(ibid., 245)

The situationists

Anarchism's theme of bypassing the state, and its rejection of modernist theory is taken further (and ultimately?) in the case of situationism – a strategy of refusing to 'play the game' of capitalist conventional society. The International Situationists were formed in 1957 and disbanded in 1972, having greatly influenced the May 1968 revolt in France:

Their writing contains a hard nosed merciless criticism of the timidity and limitations of most 'radical' opposition, including anarchism, while condemning the left, the unions and parties for their involvement in the existing order. The bottom line of situationist theory is that the greatest revolutionary idea is the decision to rebuild the entire world according to the needs of the workers' councils. . . .

(Harper 1987, 148)

The strategy for achieving this relatively conventional aim was unconventional, drawing on post-symbolist poets, Dadaists and Surrealists: revolutionary artists 'through their desire to realise and create what lies within themselves: their subjectivity'. This meant creating a world 'that has never been', which rejected the assumptions behind the 'universal structure' of capitalism 'which radiates out into every experience – culture, leisure, political organisation. In fact, all of life is dominated by the commodity and everyone participates in social life as a consumer' (p. 150).

Whereas nineteenth-century alienation was located in capitalist *production*, now it has shifted into *everyday life*, said the situationists. We are alienated from our own lives, which are simply objects to be consumed. Our lives constitute a 'spectacle' and we are the passive and powerless spectators of it. For us there is no difference between reality and its appearances, for life is a succession of images and surface appearances on which we look, but which we cannot change. There is no reality *but* them. Hence they – the appearances – *are* reality.

The relevance of this analysis, which is a postmodernist one, is heightened today, with the apparently imminent arrival of 'virtual reality'. This offers the prospect of experiencing the whole world of experiences in synthesised and 'second-hand' form: not simply via an endless round of video watching, but with smell, sight, hearing and touch through advanced computer technologies. We can become voyeurs on 'life': even watching and feeling ourselves having sex with a chosen film star:

> One side effect of technological power seems to be that human culture is growing more mechanised. At the same time human desires have been progressively stimulated, confused and ultimately numbed by the barrage of provocative images, sounds, worlds brought our way via electronic media.
>
> (Rheingold 1991)

The situationist response to all this is anarchistic, in that it refuses to accept the imposed order, encouraging each individual, existentially, to recreate and control their own lives through *détournement*. This takes elements from social stereotypes and mutates and reverses them, thereby 'exposing them as the products of alienation' and making it impossible for those who do the exposing to participate in them. We are 'to reconstruct everyday existence . . . in acts of refusal and rebellion', and particularly in 'the affirmation of pleasure and love' (Harper 1987, 152–3). These ideas were translated into the riots of the Provos and May 1968, the apparent nihilism of 1960s counterculture, hippydom and the beat generation, and student movements like the *enrages*. As Marcuse noted, some of these movements unfortunately ceased to contradict the status quo, becoming instead 'part of its healthy diet'. However, active situationist anarchism, for some analysts, was carried through into 1970s' punk culture or the 1980s' protests of the Greenham women, which were noted for a constant refusal to accept the norms and the assumptions of conventional society, including even those members of it who were 'sympathetic' to their cause (Bonnett 1989).

5

CONCLUSION: SOCIALISM AND THE ENVIRONMENT

5.1 SOCIALIST-ANARCHIST DIFFERENCES

The red-green debate

This final chapter's purpose is to lay out, in summary, the theoretical basis for eco-socialism, to suggest what ecologism should take on from Marxist analysis, and to point to the kinds of practice which are broadly consistent with eco-socialist theory. But in order to do all this we must first dispel the widespread belief that current radical green politics, which are substantially infused by anarchism, are necessarily and largely 'socialist'. This is not true, as this first section, on socialist-anarchist differences, soon makes clear. In so doing it also, of course, underlines the fact that 'anarchism' of itself is not necessarily socialist. Although, as we have noted, some commentators say it is so, it is more accurate to say that *some* anarchism has *some* common elements with socialism. But there are also many differences, and they have to be aired and understood before red-green politics can advance.

To analyse these differences is difficult, because of the need all the time to specify *what* socialists and anarchists one is talking about. For instance, the differences between anarchists and Marxists in the SPGB are not the same as those between anarchists and Marxist-Leninists. Conversely, individualist and libertarian (liberal) anarchists are far from socialism: anarcho-communists and anarcho-syndicalists are close to it. It is nonetheless worth trying to explore these differences (summarised in Table 5.1), for to do so sheds much light on the current debate among ecocentrics, between 'red-greens' and 'green-greens' (synonymous with socialistic and anarchistic perspectives respectively).

Many green anarchists, such as Carter and Bookchin, seem to equate 'Marxism', without any qualification, with state 'socialism' as it has been in East Europe, i.e. with giantism, faceless bureaucracy, tyranny and environmental degradation. Roszak equates 'socialism' *per se* with these things. But others, in condemning 'socialism' or 'communism' take care to specify that its failure in Eastern Europe is a failure of but one *kind* of socialism. Thus

to secure human freedom and environmental survival we must draw

upon socialist traditions other than that of Marxist-communism. Social anarchism is prominent among these alternative traditions. . . .
(Purchase 1990, whose 'social anarchism' is in fact anarcho-syndicalism)

Again, while some emphasise the differences between Marxism and Soviet socialism (the latter being 'the opposite to what Western socialists had been preaching' for Russell 1918, 8), others say

> surely Stalin simply took Trotsky's ideas to their extreme and Marxism to its logical conclusion. . . . Stalin . . . concludes 'by equality Marxism means, not the equalisation of personal requirements and everyday life, but the abolition of classes. . . . Marxism has never recognised, and does not recognise, any other equality'.

('RSW' 1990)

This book has argued that such a statement is substantially inaccurate, though it has not denied that there are in Marx's writings positions that could justify authoritarianism and dictatorship, as Heilbroner recognises.

However, to establish this is not necessarily to establish that there is, then, close correspondence between Marxist socialism and collectivist anarchism, as some commentators, for instance Guerin (1989), argue. The often bitter disputes between Marx and anarchists such as Stirner, Proudhon and Bakunin, were, Guerin maintains, largely the results of misunderstanding, particularly on Marx's part. Both Marxism and anarchism adopt materialist perspectives, and have the same final aim: to overthrow capitalism through class struggle.

For others, for instance Coleman (1982b), flatly deny this. Anarchists, he insists, were and are idealists, and, because they do not understand the nature of the system they oppose, they do *not* seek the same goal as Marxists. Nonetheless he also believes that 'the chief battle of socialists today is the war of ideas', and that 'Once the majority has established socialism, doing so without leaders, there will be no government and no state' (Coleman 1991), which is what anarchists believe. And Marx and Engels wanted a classless, moneyless society, without government and wage slavery: thus in this respect they, too, were 'anarchists'.

Russell (1918, 46) also drew attention to the similarity of beliefs, in communal land ownership, between socialism and anarcho-communism (i.e. 'practically all modern anarchism'). In fact, he said (p. 50), they both agree on the desired economic organisation of society but differ about political organisation and means. Something similar could be said today, about socialist and anarchist greens. There is similarity: there is correspondence, but it is not *that* close, and the gap may not be easy to jump.

Russell noted differences in their respective views on work and on distributing the fruits of labour. Under socialism a person's subsistence would come only in return for work done, or willingness to work. Anarchists

Table 5.1 Some socialist–anarchist differences

Socialism	Anarchism
Social injustice, environmental degradation caused by class exploitation	Social injustice, environmental degradation caused by hierarchical power relations
Class is defined by economic criteria	Class is also defined by non-economic criteria (race, sex)
Explanations and analyses are historical	Explanations and analyses tend to be ahistorical
Ambiguity about the state – favours at least localised forms of it	Total opposition to the state
Abolish capitalism first and the centralised state will wither away, because capitalism creates the state	Abolish the state first, as an independent act from abolishing capitalism, because the state creates capitalism
The state is the representative and defender of the bourgeoisie	The state represents its own interests, independently of other economic classes
Participation in conventional politics is permissible in the path to revolution	No participation in conventional politics is permissible
Revolution by subverting and confronting capitalism – experimental communities etc. are naive and utopian	Revolution through by-passing capitalism and creating 'prefigurations' of the desired society, such as alternative communities and economies
Emphasise strength of collective political action	Tend to emphasise personal-is-political maxim and individual lifestyle reform
Revolution particularly via our collective power as producers, i.e. unions withdrawing labour, especially in a general strike	Syndicalists advocate union organisation and action – other anarchists stress civil disobedience by community and other non-economically defined groups
Working class will be major actors in social change	New social movements and community groups will be major actors in social change
Tendency to vanguardism (Marxism–Leninism)	No revolutionary vanguards
Tendency to dictatorship of the proletariat (transitional stage)	Any 'dictatorship' or government is anathema
Materialist philosophy and approach to social analysis	Tendency to idealism
Modernist politics	Tendency to 'postmodernist' politics
Need for a planned economy	Communes should self-organise, within limits, because spontaneity is important
Limited support for decentralisation	Decentralisation is vital

Individual freedom may be circumscribed by the collective	Individual autonomy is vital
Necessary international exchange, based on reciprocity is an important aspect of international socialism	Opposed to most international trade. Applauds local self-sufficiency
Ambiguity about a money economy. Only a few oppose	Most oppose a money economy
Urban-centred	Has a prominent anti-urban element, as well as urban anarchism
Conceives of nature as socially constructed	Tends to see nature as external to society but the latter should conform to nature's laws and regard nature as a template
Anthropocentric (but not in the same way as capitalism–technocentrism)	Advocates (in social ecology) neither anthropocentrism nor biocentrism
Adovcates socialist development	Advocates various development models, inc. socialism, environmental determinism and independent development (bioregionalism)
Deep structures (especially economic) condition surface structures, such as spatial organsiation	Spatial organisation is a determinant of economics, society, politics

argue, however, for free distribution just according to need. Everyone will take what they want whether they work or not. There is no virtue in work of itself, though if it is sufficiently attractive people will want to do it.

Furthermore, while abolition of the wages system is a watchword common to both, to Russell (p. 85) '. . . in its most natural sense this watchword is one to which only the anarchists have a right' because anarchists argue that it is possible to have all life's necessities in abundance. By implication, socialists did not then argue this. Today, however, at least some socialists argue it, while, perhaps, social anarchists are now more wary about exceeding 'natural' limits to consumption as Chapters 4.2 and 4.5 intimated.

Diagnosing root causes

A major schism between anarchists and socialists, which underlies much of the differences between green and red approaches to social change, lies in their diagnoses of root causes of social injustice and environmental degradation. The current relations of production, say socialists, are those of *class*: indeed all history has been the history of class struggle. Class relations are the source of economic, social and political exploitation, and these, in turn, are what lead to ecological exploitation and damage. The true, post-revolutionary, communist society will be classless, and when it is attained the state, environmental

207

disruption, economic exploitation, war and patriarchy will all wither away, being no longer necessary.

By contrast, anarchists insist that it is *power relations* between people – relations of hierarchy and domination – which must be directly attacked. They existed before capitalism and before economic classes. Hence anarchists replace class exploitation as a diagnostic category by the more generic and ahistorical concepts of hierarchy and domination ('ahistoric', because hierarchy and domination are assumed to be possible in all modes of production, whereas socialists would argue that they were not possible in proper socialism). Bookchin strenuously rejects Marxian class analysis, and centres on hierarchy in all forms as the central problem. This is what demarcates anarchy from all other socio-political radicalism, he says (Goodway 1989).

Carter (1988, 45, 132–55) considers in detail how Marx's original two-class model has proved inadequate to describe what has happened in the twentieth century. He conveys the root of the anarchist objection to it. The problem lies in the central role given to economics by Marxist theory, and by its uncritical adoption of Hegel's view of a universal class. These mistakes meant that the possibility was overlooked that revolutionary institutions such as the party could develop political and ideological inequalities. This was no mere Stalinist aberration, for

> By viewing the proletariat as an undifferentiated minority, Marx fails to perceive the possibility that a relatively privileged labour stratum could emerge from out of those who hire out their labour power – namely a techno-bureaucracy.
>
> (ibid., 176)

This, in fact, is the ascending class today, and any future class struggle could be between it, not the 'proletariat', and the bourgeoisie. Hence, radicals must develop criteria other than economic for defining class.

These might include racism and sexism, says Jennings (1990, 207). And even though they are presently aspects of capitalist social relations they can also exist independently of capitalism. Hence if they have not been consciously and directly suppressed by any revolution, then they still could 'quite easily *emerge from our dar er sides* – even if the external material conditions now maintaining and producing them were to be disposed of'. The emphases added here suggest that Jennings views human *nature* as the ultimate source of these evils.

Coleman (1982b) rebuts the anarchists. For Marx, the materialist conception of history means that no social relations are absolute throughout time, so to object to a power relationship like oppression on grounds alone that it is always morally wrong is ahistorical. For the reason why any particular oppression occurs must always be related to a *historically specific* set of productive (class) relations, and a historically-specific cultural 'superstructure' which legitimises and otherwise sustains them. By contrast, the anarchist approach of, say, Proudhon makes 'social justice' as the pole around which all societies should evolve, the absolute yardstick by which all societies are measured. However for Marx, says Coleman, nothing can be seen as merely 'unjust' of itself – that being the end of the matter. Injustice always has to be seen in the context of, and related to, the particular mode of production under which it occurs. Thus slavery was seen as unjust in capitalism, but not in other modes of production. And it is historically naive and futile to condemn capitalism as 'unjust' of itself, for it is quite just within its own terms and premisses. (Thus, capitalists argue, logically, that freedom of opportunity does not, in the name of 'justice', have necessarily to lead to egalitarian outcomes.)

For anarchists the highest development of hierarchy and domination is in the state. Since hierarchy and injustice occur throughout history it follows that the state is also seen as a historically independent phenomenon. States arise independently of the mode of production, so abolishing the state must be an independent objective, not merely contingent on changing the mode of production from capitalism to true communism. Hence, Bakunin said that the state created capital, whereas Marx and Engels argued that capital creates the need for a state as its agent:

> Bakunin has a peculiar theory of his own, a medley of Proudhonism and communism. The chief point concerning the former is that he does not regard capital, i.e. the class antagonism between capitalists and wage workers which has arisen through social development, but the *state* as the main evil to be abolished. . . . Bakunin maintains that it is the *state* which has created capital, that the capitalist has his capital *only by grace of the state*. As, therefore, the state is the chief evil, it is above all the state which must be done away with, and then capitalism will go to blazes of itself. We, on the contrary, say: Do away with capital, the concentration of all means of production in the hands of the few, and the state will fall of itself. The difference is an essential one: without a previous social revolution the abolition of the state is nonsense; the abolition of capital *is* precisely the social revolution, and involves a change in the whole mode of production. Now then, inasmuch as to Bakunin the state is the main evil, nothing must be done which can keep the state – that is, any state, whether it be a republic, a monarchy, or anything else – alive. Hence *complete abstention from all politics.*
>
> (Engels 1872)

Carter's anarchist theory of history (Chapter 4.6) takes, it will be recalled, Bakunin's position. It argues that the state will support bourgeoisie or proletariat: whichever is most to its advantage at the time. Socialists would respond, however, that the actors for the state and bourgeoisie are by no means independent of each other. They may indeed comprise the same personnel – at different stages in their careers changing from one group to another, as for instance Hamer's (1987) analysis of the relationship between the British road lobby and the Government's Transport Department strikingly shows, or see Pringel and Spigelman's (1983) international survey of the relationships between governments and the nuclear lobby. In 1992, for example, a new chief of British Nuclear Fuels Limited was appointed: his former job was Chief Permanent Secretary in the Department of Energy (*Guardian*, 27 August 1992). In 1990, 373 Ministry of Defence officials and officers in the armed forces left to take jobs in industry, the bulk of them with arms contractors (Pallister and Norton-Taylor 1992). At the very least, most members of the state apparatus have come from the same exclusive educational, social and economic backgrounds as the bourgeoisie, thus acquiring closely coincident world views. (Both Sampson (1984) and Paxman (1991) remark on changes in power relations between government and bourgeoise in Britain, such that a 'new set of entrepreneurial figures who had made it on their own terms' and had no common background, education or social network enjoyed access to power in the 1980s (Paxman). But they also acknowledge the resilience of the old Establishment: 'alongside this new meritocracy there still remains a remarkable educational elite which has maintained its continuity and influence through all the political upheavals' (Sampson). From Eton and Winchester, this elite constituted a mix of civil service bureaucrats – an immobile bureaucracy which can, as Carter suggests, perpetuate its own interests and values – and entrepreneurial aristocrats.)

Again, this anarchist–socialist disagreement is much reflected in the debate between 'green dreams' and 'red realities', where Ashton (1985, 21) tells greens that government is closely tied into the objectives of capitalism, hence the belief that government will

> take on board the demands of the Green movement [not only] . . . amounts to a denial of the rationality, within capitalism, of environmentally irrational production . . . [but also] incorrectly assumes government to be independent of the powerful vested interests which help determine the nature and perpetuate the rationale of capitalist production.

In support, Ashton describes how America's Environmental Protection Agency in the 1980s was 'stacked with industrial lobbyists and lawyers whose aim was to loosen environmental control'. Indeed, this was just part of a larger process in which a group of industrialists bought Ronald Reagan into the presidency to act as their agent, partly by systematically undoing the fetters on industry which had resulted from the environmental legislation passed in the 1970s (Faber and O'Connor 1989). (See also the example of the American government's defence of business interests against impending environmental agreements at the Earth Summit in Chapter 3.4.)

Strategies for revolutionary change

Anarchists and socialists both want revolutionary change. But whereas anarchists reject the political process and advocate 'direct action' socialists often seem prepared to follow a dual philosophy. On the one hand, withdrawal of labour, especially in the general strike, is a form of direct action time-honoured by the labour movement, which Morris, in common with anarcho-syndicalists of the day, saw as the linchpin of revolution. But on the other hand socialists also tend to accommodate with the state in its form of parliamentary 'democracy', thereby drawing anarchist contempt.

Such accommodation ranges from the 'managing capitalism' approach of social democracy and Fabianism to the attempt to use the parliamentary process by groups who embrace Marxism's ultimate antipathy to the state. Thus *Militant*, in Britain, wants the Labour Party to adopt its ideas, policies and programme, believing that:

> The combination of a mass based and campaigning Labour Party with a Marxist programme would make the labour movement invincible.
>
> (Taffe 1990, 28)

And the SPGB, while agreeing in spirit with Morris's view of Parliament as potentially more useful as a dung market, nonetheless intends to use it, and the mass vote of the working class, as a revolutionary instrument (see Chapter 3.8).

The forms of 'direct action' which anarchists espouse span personal lifestyle change, setting up communes and coops, strikes and boycotts, non-violent demonstrations, refusals and obstructions and violence against property but not people. Some, like Carter (1990), advise attack on all fronts, including non-violent direct action 'to disempower the state'. More commonly anarchists tend to split into camps, depending, for instance, on how much they accept or reject notions of class war and collectivity, and/or how much they embrace pacifism.

Cook (1990) has noted how the late nineteenth-century anarchists' emphasis on 'propaganda by the deed' was partly responsible for Marxists rejecting anarchism. For one thing, it is ultimately *counter*-revolutionary, as Coleman (1982b) suggests: Baader-Meinhof and other self-styled 'anarchistic' terrorist groups merely encouraged people to stand by the state and its police and armed forces. Cook (p. 17) detects, however, that anarchists cannot win in this debate

> Whereas Proudhon's motto was *destruam et aedificabo* (I destroy and I build up), Bakunin and his followers became associated with the destruction side of the equation and in the latter years of the century anarchists turned increasingly to individual acts of terrorism against the established order. It is this era and the images associated with it which have done most to weaken anarchism's popular appeal. Those anarchists who rejected violence as a means to an end were few in number; even

Kropotkin with his gentle personality remained ambivalent about violence and lost much credibility when he supported the First World War with Germany. Conversely, and perhaps paradoxically, pacifist anarchism is also the subject of fierce criticism from those who see non-violence as an inadequate means of attaining revolutionary objectives.

The latter may include Marxists, like Morris. Even if, they argue, the proletariat does not desire initial violence in seizing the centres of power, capital will use it in attempting to put rebellion down, and so workers must be prepared to fight to defend their lives and their cause.

Allied to this issue is the argument over spontaneity and *mass* revolution, as opposed to vanguardism. 'All communistically oriented anarchists are characterised by a belief in the spontaneous character of social revolution' says Sekelj (1990, 20), and

> an anarchist revolution is one of generalised rebellion, without leaders and masses. Each social unit shakes off the fetters and mechanisms imposed on it by political and economic powers.
>
> (Baldelli 1971, 24)

Spontaneity is a key principle of anarchist society, as is non-hierarchy. Therefore, according to the prefiguring principle spontaneity and non-hierarchy must guide all revolutionary phases. Leninist socialists like the Socialist Workers' Party often counter that this is naive and idealistic. For one thing, a spontaneous outbreak of revolutionary consciousness among the majority of people is precluded in capitalism by the hegemony which capital exercises over the media, education and socialisation. As Orwell noted, people will not rebel until they have consciousness, but they will not have consciousness until they rebel. Hence a vanguard of people with a clear theoretical analysis, a practical strategy for revolution, and a preconceived organisation is essential (Chapter 3.8). Furthermore:

> A revolution is certainly the most authoritarian thing there is; it is the act whereby one part of the population imposes its will on the other part by means of rifles, bayonets and cannon – authoritarian means if such there be at all; and if the victorious party does not wish to have fought in vain, it must maintain this rule by means of the terror which its arms inspire in the reactionaries.
>
> (Engels, cited in Joll 1979)

Not all Marxists take this view. Coleman (1982b) argues that Marx saw revolution coming only from fully conscious socialists taking power as a majority, and he repudiated vanguardism. This is the SPGB's position, too. Its eventual success will come when its present role as an intellectual vanguard, proselytising socialist consciousness, becomes redundant as a result of the majority of people taking on its ideas. In the ensuing revolution, the SPGB will be able to disband itself and fade into the general revolutionary masses. Leading on from this, Coleman also maintains that Marx was not advocating a 'workers' state' as Lenin argued, but that to get power workers would have to overcome the state, use it to dismantle coercive forces and eventually create a society where the working class and state would dissolve themselves. Thus the SPGB dissociates itself from aspects of Marxism which appear to derive a model of a 'higher stage' of communism, preceded by a transitional 'lower', socialist, stage involving, in essence, dictatorship of the proletariat. In so doing it aligns itself with anarchism's aversion to hierarchy during any revolutionary stage. Here, the anarchist critique of

Marx, Social Democracy and Bolshevism is very fertile and empirically confirmed. Its essence can be reduced to Bakunin's critique of Marx – that there cannot exist a dictatorship of a transitory character, least of all a dictatorship of the proletariat. The state, both democratic and dictatorial, cannot be the means of the realisation of communism, and this anarchistic objection equally applies to Social Democracy and Bolshevism. History has shown that freedom cannot be realised through dictatorship, a free community of associated producers through the state, and also that parliament cannot be a substitute for social revolution.

(Sekelj 1990, 20)

It seems that, at best, Marx was 'ambiguous' about the state, as Russell put it, or 'evading the issue' as Guerin (1989, 121) says: 'Anarchists suspect Marxists with good reason for not having purged themselves completely of any Jacobin inclination'.

Guerin here refers to the two French revolutions. One, the Jacobinite, was a revolution from the top by the bourgeoisie and was dictatorial in style. The other was libertarian, communalist and by a 'proto-proletariat' of small artisans. The latter culminated in the 1871 Paris commune and the 1917 Russian soviets. Marx and Engels swung perpetually between the two, and between state nationalisation and workers' self-management of the means of production. In the 1848 *Communist Manifesto* they wanted the first, but the preface of the 1872 edition concedes to anarchism, calling for 'self-government of the producers'. Rosa Luxembourg, says Guerin, was a link between anarchists and 'genuine Marxists', with her call for the mass strike and socialism powered from the bottom up by revolutionary councils.

The traditional socialist view has held the working class to be a definable entity, who would be principal actors in revolution, whether or not there was also any vanguard. Still, today, this is a common socialist position, as is the perceived need for a continuing state, but one which works more in proletarian interests. Socialist critics of postmodernism and of the green movement, such as Frankel, Ryle and Ashton, maintain the centrality of the labour movement in social change, even if they do not go as far as some who might believe that

> a new layer of the working class, not yet in positions of influence either in the trade union branches or even in the shop stewards' committees is coming to the fore. This new layer is destined to play a decisive role in swinging the trade unions and the Labour Party to the left.
>
> (Taffe 1990, 32)

By contrast, green anarchists today, like Bookchin, have largely forsaken the syndicalist strand in their history, instead pinning faith in new social movements as principal actors (Chapter 3.8), and pouring scorn on 'worker worship'. They would take their cue from Bernari (1990), who as long ago as 1936 dismissed the proletariat as a revolutionary force. He saw them instead as ignorant of themselves, lacking class consciousness, unintelligent and weighed down with prejudice, infantile illusions and false consciousness manifesting itself in racism and support for war industries. Hence workers fight for their own long-term interests only reluctantly. The continuing truth in this anarchist evaluation of the working class is brought home when one considers where support for today's right-wing gutter media, like the *Sun* newspaper, and groups like the National Front mainly comes from.

Green anarchists scorn the socialist path and labour actors, drawing instead on anarchism's tradition of bypassing the state and setting up exemplary decentralised communes and groups devoted to radically different lifestyles. As was the case a century ago, when it was the less well off artisans and rural labourers who followed this essentially utopian socialist path because they had little to gain from industrial society (see Goodwin and Taylor 1982 and Gould 1988), today the actors in this green anarchist movement are often hostile to industrialism, technology and modernisation, and shun both capital and labour. They draw criticism from socialist greens, echoing the original Marxist critique of utopian socialism (Chapter 3.9):

> The 'exemplary project' and 'moral stand' have a legitimate place in any strategy for change . . . [but] As a wholistic strategy for change in a political economy dominated by powerful vested interests . . . it is unlikely however to provide an adequate momentum for change. . . . Despite pacifism people do not all refuse to go to war, despite feminism most people still live in families and despite the growth of cooperatives most people still stay in more conventional employment or on the dole.
>
> (Ashton 1985, 20)

Indeed, though socialists also set up cooperatives, these are distinguished from the small and specialist green coops by their seeking the backing of organised labour, which greens distrust.

The need for a state

Besides divisions over how necessary it is for the state to figure in the revolutionary process, there are also differences between anarchists and socialists over the need for a state in the post-revolutionary society. Anarchist–communists continue to declare that the state is wholly inconsistent with their philosophy, even though, as Sekelj (1990, 21) points out

> At the very moment that anarcho-communists as a movement represented a serious political, social and armed power at the time of the Spanish Civil War, they themselves transgressed this principle

by joining the government and accepting compromises.

Sekelj questions the principle, distinguishing between a need for libertarianism in the economic sphere to encourage a rich tradition of workers' self-management, and a need for a state to regulate in the political sphere to secure the egalitarian representation of all citizens. (We have already noted how Russell also thought this necessary because of a 'natural' human desire to tyrannise others.)

For the anarchist vision of direct democracy, based on Rousseau's conception that any form of representation is an alienation of freedom, is probably only possible in a homogeneous society, where everyone conforms to a common world view. Without this, any 'democracy' through mandating delegates could become almost a totalitarian caricature. Hence Sekelj (p. 24) concludes that

> The state and bureaucracy, even under the condition of an option for a different type of technology and economic growth, are unavoidable factors of the organisation of a society on that level of complexity at which it exists today.

The great socialist-communist projects of the nineteenth century, like abolishing the state, should, he thinks, be revised, and socialism must meet social reality. For today's radical experimenters will never be able to achieve the original aspirations of anarcho-communism or Marx's communist utopia. The 'most significant' among these are the kibbutzim, but they cannot, and indeed do not want to 'transcend the main current of . . . global society' (p. 23).

Like Sekelj, O'Connor (1991a, 27) believes it important to try to *democratise* the state rather than to attempt Bookchin's social ecology project of abolishing it. He considers (1991b, 34–6), that most economic, social and ecological problems

> cannot be adequately addressed at the local level. . . . Regional, national and international planning are necessary. . . . Furthermore, if we broaden the concept of ecology to include urban environments, or what Marx called 'general communal conditions of production', problems of urban transport and congestion; high rents and housing; drugs and so on, what appear to be local issues amenable to local solutions turn out, in

Table 5.2 Frankel's semi-autarky economy

CENTRALISED PLANNING

To administer
taxation, currency controls, international trade, post and telecommunications, budget coordination, guaranteed minimum income scheme, and so forth.

To plan
use of raw materials, energy and fuel, and so on.

To coordinate
manufacturing industry (e.g. machine tools, heavy equipment, steel production).

The state provides capital,
raw material
social income to

DECENTRALISED LOCAL AND REGIONAL SECTORS

Sector 1 Employment in state institutions – health, education, transport, etc. – steered by public debate, not 'markets'.

Sector 2 Self-organised worker coops for local and regional needs. Partly state and partly market controlled (the former constrains the influence of the latter).

Sector 3 Individual and family producers, not subject to planning but:
(a) regulated by a market (food, crafts, personal services, repairs)
(b) 'prosuming' or simple barter, moneyless, self-sufficient transactions.

---

fact, to be global issues pertaining to, e.g., the way that money capital is allocated world wide. . . .

Hence the socialist argument against anarchism's utopia and for a state becomes an argument against autarky (decentralisation, self-sufficiency and local autonomy) and against an allegedly unrealistic refusal to accept the complexities of a technological world as it is. It is partly pragmatic, pointing out the difficulties of organising from the bottom up.

Frankel (1987) presents a catalogue of such difficulties in repudiating both right-wing and left–anarchist autarky. For instance, greater regional autonomy will require interregional and international bodies to mediate between regional interests; (extended) families cannot be responsible for curing all social ills, therefore law and regulation will be needed to protect against such things as racism or child abuse or quarrels over personal possessions and the like; state institutions guaranteeing incomes and local resources for education and culture are needed to prevent a drift towards primitive parochialism. To conserve energy, economise on transport, to have wages, social security and money (which will continue to be needed for the foreseeable future) and participatory democracy 'Then social planning and open public discourse become essential' (p. 252).

Frankel's preferred vision of 'semi-autarky' (Table 5.2), which the eco-socialist Ryle (1988, 85) also approves of, hinges on the notion of democratising the state. It is another mixed economy scenario, in a way like the Gorz schema for heteronomous and autonomous spheres which Frankel criticises. The state functions at both national and local and decentralised levels, and there is also a sector of individual and family producers, unplanned, and regulated only by a market. As such, the socio-economic visions of the likes of Gorz, Frankel, Ryle and O'Connor are no longer founded on truly radical socialism of the William Morris kind, without wage slavery or automation or money or the state. To find such fundamental socialist visions today, one must turn to the SPGB or some anarchists.

But this is not to say that such 'revisionist' socialist views are unattractive. While most elements of Frankel's semi-autarkic economy exist in today's social democrat programmes, the relative emphasis away from market control makes Frankel's vision a more radical one. And, again, it is an open question whether 'the state' of itself need be the unresponsive, oppressive and over-bureaucratic institution which it is popularly supposed to be among market liberals and radical liberal–leftist new social movements alike. The state in Britain, containing many dedicated, highly efficient, responsive, caring, unbureaucratic and socially conscious people, has been subjected to a deliberate, sustained and successful programme of vilification by the New Right, in which it has effortlessly and falsely been equated with Soviet bureaucracy and Orwellian totalitarianism. In fact, as Frankel (1987, 202–3) suggests, greens, anarchists and post-industrial theorists have one-sided and simplistic views of 'the state'. It is not a mere administrative machine existing outside 'the economy' or 'civil society' – or us. There is a complex interpenetration of all three sectors of society and we, individually, are intimately part of it.

This does not have to be for the worst, as Ryle (1988, 256) affirms: 'There is no inherent quality in central planning which makes it undemocratic'. And the greater democratisation of the state is not an unrealistic expectation, though it will have to be substantially based on decentralisation. Nor, if this were to be achieved, does it appear axiomatic that largely locally-based, community-accessible state agencies would necessarily be any more hierarchical or oppressive than the regional federal bodies which anarchists all acknowledge would be needed in their ideal society. Very much will depend on the relative success of the built-in democratic mechanisms in state and federal structures. If, as Kohr (1957) maintains, the USA is an example of a 'successful' federation, then the issue is in doubt, for the USA is a gross perversion of virtually everything in the socialist and anarchist canons.

5.2 PUSHING ECOLOGISM CLOSER TO ECO-SOCIALISM

Ecologism currently owes much to anarchism, and is also infused by deep ecology and new ageism. At the same time, many greens mistrust or renounce what they take to be Marxism and socialism. I believe that such thinking should be reversed, bringing Marxist analysis much more into ecologism's mainstream, and shedding liberal facets of its anarchism, in favour of more communist and syndicalist–anarchist traditions. To say this is not necessarily to argue that ecologism should take on *Marx*, lock, stock and barrel. For, as has been revealed, there is no one interpretation of Marx, and I cannot say which one is most 'correct'. But this does not matter: neither does it matter if Marx was some sort of covert ecocentric (Parsons) or was not and never would be (Grundmann). What matters is that to look at the political questions outlined in Chapter 1 from the perspectives of many of the Marxists reviewed in Chapter 3 yields valuable insights that might make ecologism a more coherent, potent and appealing ideology – which must ultimately be a form of socialism. To achieve this, however, anarchist and deep ecology tendencies towards liberal political economy and ideology, anti-humanism, mystification and idealism must be shed.

Marxism's perspectives on the issues raised in Chapter 1 would involve (1) seeing human nature as largely socially constructed and therefore changeable – yet emphasising basic human needs for communality and production; (2) opposing crude determinism, materialism and economism, but also opposing idealism. Humanity is progressively freed in evolving to communism, but the ultimate constraints of nature and historical legacy are acknowledged and we can never have total free will to construct what is 'out there'; (3) taking a *dialectical* materialist approach to history and social change, which acknowledges the importance of ideas, subjectivity and spirituality, but also relates them to economic contexts and always rejects mystification – be it of economic, social, spiritual or nature forces; (4) aiming, in communism, for ultimate fulfilment for individuals, but emphasising collective approaches to social change; (5) seeing communism as a secular *gemeinschaft*; (6) regarding latent class (particularly economic) conflict as a potent force in shaping society and history; (7) having a structuralist perspective which particularly thinks about how surface appearances manifest underlying economic class relations; (8) advocating the specific development model of socialism/communism, predicated absolutely on (9) egalitarianism and rejecting the market as regulator of economic, political and social behaviour. Eco-socialist politics would also embrace modernism and fight for absolute humanist values like individual freedom and fulfilment and egalitarianism, informed by rationality as the main criterion for judgement. At the same time they would understand absolutely how morality is culturally and historically constructed. And eco-socialist political economy would be based on the abstract labour theory of value, whether it proposed money in the economy or not.

The need for socialism and class action

Sustainable, ecologically sound capitalist development is a contradiction in terms. For reasons outlined in Chapters 3.3 and 3.4, capitalism is growth-oriented – growth in real values resting on the exploitation of nature, including human labour, in production: and it is necessarily technologically and organisationally dynamic:

> This implies that capitalism has to prepare the ground for, and actually achieve an expansion of growth in real values no matter what the social, political, geopolitical or ecological consequences.
>
> (Harvey 1990, 180)

Contrary to many green and non-green claims, this growth dynamic is not negotiable for environmental or social justice outcomes. As President Bush said, when refusing to sign the Earth Summit biodiversity accord in 1992: 'We cannot permit the extreme in the environmental movement to shut down the United States' (i.e. world capitalism).

Crisis in capitalism is endemic, and is defined by the system as *lack* of growth: but that, like its opposite, is bad news for greens. For during recession industry can and does more nakedly resist and erode environmental protection regulation in the interest of 'national' economic interest. It undermines the strength of worker opposition to economic and ecological exploitation. And, during recession people's minds turn from issues other than economic immediacy. Poverty is still the greatest foe of liberty, says Galbraith (1991), and freedom from want bears decisively on our desire for other freedoms – of speech and expression, of worship and from fear. We should acknowledge this reminder of the need to appreciate economic determinism, and add to Galbraith's trilogy that freedom from pollution and environmental degradation also requires a measure of material wellbeing. So, too, does the development of ecological consciousness. And an ecological–communist utopia requires the development of productive forces. To say this is not to accept the fatuous market liberal argument that economic growth (of any kind) is needed to 'create' the wealth required to be able to afford to clean up the environment (i.e. to clean up the mess created by the growth in the first place).

Eco-socialist growth must be a rational, planned development for everyone's equal benefit, which would therefore be ecologically benign:

> A society based on common ownership and democratic control, with production solely for use not sale and profit, alone provides the framework within which humans can meet their needs in ecologically acceptable ways.
>
> (SPGB 1990, 2)

Such socialist development *can* be green, being predicated on the maxim that there are natural limits to every human's material needs. They are needs which can therefore be met within the broad limits of nature's ability to contribute to productive forces. The fact that in socialist development people continuously develop their needs to more sophisticated levels does not have to infringe this maxim. A society richer in the arts, where people eat more varied and cleverly prepared food, use more artfully constructed technology, are more educated, have more varied leisure pursuits, travel more, have more fulfilling relationships and so on, would likely demand less, rather than more, of earth's carrying capacity, as any green will tell you.

It follows that the best green strategies are those designed to overturn capitalism and establish socialism/communism. Anarchistic prefiguring strategies that attempt to marginalise capitalism sound seductive, but experience has shown that they usually result in the marginalising of the countercultural marginalisers themselves, who have ignored or underestimated or refused to confront the material basis of the hegemony of prevailing capitalist ideologies.

But so many ecotopias – those of Gorz (1982), Callenbach (1978, 1981) and Piercey (1979) for instance, are really liberal (therefore capitalist) dreams – emphasising *individual* creativity as the ultimate fulfilment (Byrne 1985). They reject collective social change strategies, such as syndicalism and other trades union activity, like combined committees' workers' plans, which '. . . specifically address the problems posed by capital's reorganisation in the face of a [previous] working class offensive' (Byrne, p. 82).

Bookchin's (1986b) anarchism similarly rejects syndicalism and unionism as narrow, class-based, and the product of anarchist and socialist Germans, Italians and Jews who came to America and were confined to ghettoes. He instead champions the struggle of Anglo-Saxon society for rights that express higher ethical and political aspirations than the 'myth' of a workers' party, and which reach highest development in the American Congregationalist town-meeting where middle and working classes join in one people's movement. Purchase (1992) is outraged by such apparently shallow populism, in pursuit, he says, of Bookchin's desire to be the sole guru of 'social ecology'. Bookchin here has sweepingly dismissed centuries of resistance to capital by ordinary workers, who have now linked this struggle to the green concerns of Australian communities via the 1970s' 'green bans' (see below).

Purchase correctly asserts that in the absence of capitalism

> trades unions and workers' cooperatives . . . would seem to be a quite natural – indeed logical and rational way of enabling ordinary working people to coordinate the economic and industrial life of the city for the benefit of themselves. [Furthermore] Ecoactivists . . . have yet to digest the hard historical fact that the institution of state-sponsored multinational exploration cannot be defeated without the commitment of large sections of the organised working class to the green cause.

Some, but not all, anarchists want to emphasise the socialist notion of collective class struggle, and indeed to deny, like Rooum (1992), that anarchism has any other orientation: individualism and class struggle being two facets of the same thing. However, eco-anarchists and mainstream greens usually are among those who discount the continued existence or revolutionary potential of a working-class proletariat. Yet the alternative agents, suggested by neo-Marxists or others, such as the unemployed or new social movements, are even less convincing 'revolutionaries' – the latter, again, being empathetic with liberal rather than socialist traditions and ideals.

Yet there is no *a priori* reason why ecologism should be linked to political conservatism or liberalism, rather than the labour movement and socialism. Indeed, some insist that there is inherently a strong left–green affinity (e.g. Steward 1989, Dobson 1990). Though red–green coalitions are problematic, for reasons suggested in Chapter 5.1, they do become more feasible when contemplated between libertarian socialists like Byrne and green anarcho-syndicalists like Purchase. For they both recognise the potentially vital role of class analysis and struggle, and base their ecologism on a wide definition of 'the environment', drawing on what is, after all, a traditional trades union concern for improving environmental quality of life for the low-paid urban masses. Socialists like Robert Blatchford, and unions like NALGO, were in the 1920s and 1930s heavily involved in back-to-the-land and holiday camp movements, for instance (Ward and Hardy 1986, Gould 1988).

Defending anthropocentrism

Marxism encourages us to formulate an ecologism that espouses much of the Enlightenment project, defying the green postmodernism of, for instance, Atkinson (1991, 43–4, 61–2). It wants to secure the material welfare of all humanity, through growth of productive forces via the 'domination' of nature. But it rejects modern industrialisation (capitalist or East European 'socialist') which, 'masters' nature by transforming it to the detriment, in the broadest senses, of humans. Marxism has often been at fault for not clarifying that it sees a difference between 'mastery', which implies subjugation or destruction and 'domination', which does not: Marx's 'domination' does not cause ecological problems, says Grundmann (1991, 15) but is the starting point from which to address them: ' "Domination of nature" is not responsible for ecological problems; quite the contrary: the very presence of ecological problems proves the absence of such domination'. Here, 'domination' means *collective conscious control by humans of their relationship with nature*. This implies stewardship rather than destruction (see Attfield 1983). It also implies socialist rationality and humanism, recognising the folly and injustice in creating harmful and unpleasant environments, anywhere. But it asserts that true human freedom is possible only in 'second nature':

> The more first nature is transformed into second nature, the more its laws are understood and the more mankind is able to free itself from its strains. Communism is the culmination of such a process . . . only a society which is able to control its own workings on the natural environment is worth the name communist.
>
> (Grundmann, p. 11)

221

Ecocentrics would criticise this view for being 'anthropocentric' rather than 'biocentric'. And it *is* anthropocentric in the sense that its concern for the state of 'nature', which it sees as largely socially produced, is triggered by the traditional humanist concerns of socialism. Therefore it cannot consider 'nature's needs' aside from those of humanity, and just as it considers that a communist society cannot, by definition, be ecologically unsound, so it would assert that a proper ecological society could not countenance, by definition, social injustice. It would also prioritise human over non-human needs when the chips are down.

However, this kind of anthropocentrism is not equivalent to that of capitalist technocentrism, since it is rooted in the concept of a society–nature dialectic, with all that this implies (Chapter 3.6). Bookchin considers that a proper perspective on this dialectic would result in no 'centricity' at all – neither bio- nor anthropo-. But this does not necessarily follow. An even-handed appreciation of how society fashions nature and nature society, and of the desirability of being ecologically benign, does not mean that humans should be even-handed about their position as members of the *human* society. Within the broader understanding afforded by dialectics of the intimateness of our ties with the rest of nature it is quite legitimate to prefer those things that advance the interests of humanity and do not destroy the rest of nature (even though they must inevitably continue to refashion – 'produce' – nature). In just the same way it is legitimate for elephants to prefer what is in the interests of elephants.

So eco-socialist anthropocentrism is a long-term collective anthropocentrism, not the short-term individualist anthropocentrism of neoclassical economics. It will therefore act to achieve sustainable development, both for pragmatic material reasons and because it wants to value nature in non-material ways. But ultimately this latter will be for the sake of *human* spiritual welfare.

Most socialists also make the obvious point that biocentrism is not actually possible for *humans* to achieve. Ecocentrism

> pretends to define ecological problems purely from the standpoint of nature. . . . But it is obvious that the definition of nature and ecological balance is a *human* act, a human definition . . . in relation to man's needs, pleasures and desires.
>
> (Grundmann, p. 20)

Similarly, says Watson (1983), a preference for 'climax' biomes, variety, balance, equilibrium and complexity is a matter of *human* economics and aesthetics. We do not know that 'nature' prefers these things any more than it recognises the notion of injury to others (indeed, Lovelock tells us that the earth's atmosphere today can *not* be regarded as being in equilibrium, even when the effect of industrial society is discounted).

To assert all this is not the mere obvious sophistry that Eckersley (1992),

Fox (1990) and other biocentrics maintain. For to assert the reverse is to maintain that there are objectively-existing 'natural' values – out there – to which we must submit (Commoner's 'nature knows best' principle). Callicott (1985), a defender of deep ecology, readily concedes that notions of 'intrinsic' value do imply objectively existing value, and rest ultimately on the Cartesian dualism. This is deep ecology's paradoxical dualism despite its ostensive claim to monism, which I referred to in discussing alienation (Chapter 3.6). Callicott, in fact, turns away from it, recognising that intrinsic value cannot be logically equated with some objective natural property, independent of any subjective or conscious preference. This seems wise, for intrinsic value theory can lead us down a path which views

> nature as being natural, undisturbed and unperturbed only when human beings are *not* present, or only when human beings are curbing their natural behaviour, [and] then we are assuming that human beings are apart from, separate from, different from, removed from or above nature. . . . To avoid this separation of man from nature, this special treatment of human beings as other than nature, we must stress that man's works (yes, including H-bombs and gas chambers) are as natural as those of bower birds and beavers
>
> (Watson 1983, 252)

Thus only humans are not allowed by deep ecologists to be natural; all other species are morally neutral, but humans are morally bad, even though

> Man's nature, his role, his forte, his glory has been to propagate and thrive at the expense of many other species, and to the disruption – or, neutrally, to the change – of the planet's ecology . . . human beings do alter things . . . this is their destiny.

Indeed, the universe itself resulted from explosive, disruptive change. In the event, this is also Gaia theory's message:

> Organisms are adapting in a world whose material state is determined by the activities of their neighbours, this means that changing the environment is part of the game.
>
> (Lovelock 1989, 33)

The evolution of species and of their environment is 'tightly coupled' (or could one say 'dialectically related'?). So in a world where, as Lovelock thinks, mountain range formation and tectonic plate movement was initiated through the activity of microorganisms, it seems paradoxical to see 'man', universally, as the great sinful destroyer.

The deep ecology tendency, when unmoderated, to do just this does contain contradictions. Eckersley (1992, 53) claims, after Fox (1990) that all creatures should be allowed to develop in their own way. But she does not apply this to humans, or, alternatively, she thinks that transforming nature through

223

industry is not humankind's own way, i.e. it is not intrinsic or natural – in which case it has to be determined why humans have behaved unnaturally for the past few hundred years at least. Even more confusingly, Eckersley goes on to concede (p. 156) that change, innovation, destruction and extinction *are* natural, but, conservatively, 'must be allowed to unfold in accordance with natural successional and evolutionary time'. Yet she also does not claim to know what the thrust or direction of evolution is, consequently it is unclear that what humans now do to nature is deviating from natural evolution as she implies.

Watson, having analysed the writings of Naess, Sessions, Rodman and others, describes this deep ecology position as 'anti-anthropocentric biocentrism' (ecosophy). Human desires are not privileged – humans are regarded as an equal part of a holistic system – and they should not change the planet's ecology: the world ecological system is too complex for humans to understand, and the ultimate human goal and joy is to contemplate, not change, nature, drawing sermons from the stones.

This is all very different from the Marxist socialist, monist, anthropocentric position of egalitarian development and growth through human labour and scientific ingenuity, to satisfy materially limited but ever richer human needs through democratic, collective, planned production that emphasises resource conservation, non-pollution, recycling and quality landscapes. Despite some claims to the contrary, there is no possibility of a 'socialist biocentrism' (O'Connor and Orton 1991), since socialism by definition starts from concern over the plight of humans. Contrastingly, what distinguishes ecocentrism (according to Eckersley, p. 28) is, *first*, what is the 'proper' place of humans in nature? and, *second*, what are the social and political arrangements appropriate to this place: the first determines the second.

As noted above, there are still neo-Marxist influenced writers, like Bookchin, who claim that the Marxist 'dialectic' undesirably overemphasises humans. Bookchin therefore calls for a more 'naturalistic' dialectic opposed to both deep ecology and Marxism. In practice it is difficult to see what would be different in it from the socialist position outlined above, except for more emphasis on human spontaneity and less on planning and control of nature. For socialist anthropocentrism does *not* do what Eckersley (p. 50) accuses conventional anthropocentrism of: it does not justify moral priority for humans on the basis of their *separation* from the natural world. Neither does it logically follow, as she asserts (p. 52) that humans are thereby justified in plundering the natural world or holding non-human nature as valueless.

These differences, between social ecology (green–postmodernism), deep ecological and Marxist conceptions of the society–nature relationship are neatly encapsulated by O'Connor (1992 pers. comm.):

Three terms, culture, nature, social labor.

If you think that culture mediates nature and social labor, you're a post-Marxist, post-modernist, deconstructionist, etc., i.e. idealist.

If you think that nature mediates or brings together social labor and culture, you're an environmental determinist, bioregionalist, deep ecologist, essentialist, eco-feminist, etc., i.e. a passive materialist.

If you think as I do that social labor mediates culture and nature (which in turn mediate back in various ways), you're an eco-Marxist. An active materialist.

The state, autarky, planning and money

Green socialist development should show the following:

Social relations and socio-economic objectives prevailing in the past must be reversed . . . class and gender relations must be egalitarian rather than unequal, productive resources must serve local people rather than distant demands, and decisions must be democratically made rather than the prerogative of the elites.

(Peet 1991, 165)

This implies acceptance in large measure of the abstract labour theory of value, because subjective preference and cost of production theories are based on the proposition that a large part of exchange should be regulated by the logic of a 'free' market (even if intervened in). Both radical green and radical socialist perspectives, however, inexorably lead to a position where, although markets might exist, they cannot be allowed to occupy their present prominent position as the regulator of people's lives – indeed the nexus through which all life has to be mediated. It follows that there should be withdrawal from world capitalism, which cannot properly meet needs in the periphery after satiating demands in the centre, because trickle-down theory does not work. And there must be many cooperative institutions matched with local conditions under democracy and mass participation, and industrialisation through indigenous resources and technology.

Is this socialist model the same as that of green, anarchist post-industrial autarky? Does it also mean the end of the state – even more, of markets and money? Marx himself was ambiguous about the latter questions, according to Grundmann (1991, 265–71). He argued that a state is necessary and can be used by the proletariat to create communism, but also that it must and will wither away, to be replaced by networks of relatively autonomous producers. Furthermore, he seemed to advocate variously both markets and/or a central plan, and both money and/or certificates of labour quanta as means of converting concrete to 'abstract' labour (a universal unit of value).

Modern socialists also have no one view. Frankel, for instance, like Sayer (1992), advocates a mix of markets and state planning, with money existing for the foreseeable future. Anderson (1991, 46) believes that an 'enlarged

225

economics – a renewed political economy' (essentially socialistic) can be achieved in conventional terms as long as it incorporates new kinds of economic indicators. These would have three parallel ways of describing economic processes, considering the economy from a monetary or financial point of view and simultaneously as consisting of human beings organised together in particular ways and as a set of arrangements for mediating the relationship between humans and the natural world. At least in the short term, he says (p. 95) there should be changes within the existing framework of national income accounting, especially for the addition of figures for the value of unpaid work and the subtraction of figures for environmental depreciation. However, Buick and Crump (1986), who anarchistically regard the state as oppressive *by definition*, elaborate some details of the moneyless economy which they regard as essential to avoid appropriation of surplus value. Theirs is a very logical position deriving directly from Marx's analysis of the alienation and commodity fetishism that follows a money economy (see Harvey 1990, 100–3).

But one thing seems clear. Changing spatial organisation does not *of itself* create green socialist development. If it did, then the market liberals and post-industrial theorists might not be as keen on elements of autarky as they apparently are. Some (e.g. Sabel 1982) have eulogised developments like those in the 'third Italy' (between industrial north and underdeveloped south). Here, small firms (e.g. engineering) have apparently been successful and autonomous, have collaborated across skills and management levels, and have resurrected craft production, in an apparently post-Fordist autarkic network.

However, Murray (1987) shows the reality behind this small-scale 'green' utopia, serving diversified world markets and consumer tastes. It is, in fact, absolutely part of, and dependent on, centralised *Fordist* production, for it thrives principally by supplying the needs of multinationals subcontracting their work in a regime of flexible accumulation. Working conditions are atrocious, work is uncertain, and labour, being fragmented, is powerless to help itself. Here, autarky makes problems *worse* because it occurs within a capitalist economy.

And there are many other objections to autarky, which Frankel details (see Chapter 5.1): there might be increased isolation and parochialism in the electronic cottages, basic communes or bioregions with no mediating institutions between them and the global order (see also Eckersley 1992).

Such difficulties highlight a fundamental problem shared by anarchism and decentralist libertarian socialism. Their autarkic, fragmented and confederational 'utopias' can only work if most citizens share a set of fundamental beliefs. But this would contravene another fundamental belief: that *differences* and individualism should be allowed, if not celebrated. Socialist democracy must not lead to what Orwell called the tyranny of the majority. Here anarchists and socialists are thrown back onto believing in a basically communal human nature, which will out in communism – with all the attendant problems of such reasoning (Chapter 4.3), not least of which is most people's scepticism.

Grundmann (1991, 293) evokes Marx's own arguments in *Capital* against 'simple models of social life which . . . take as their reference point the community and autarky of the Greek polis'. They cannot realistically reverse what has been a hugely strong historical (even ahistorical) drive towards functional differentiation – spatially and between economic, political and technological subsystems – creating a highly mobile and pluralist world society where traditional bonds are irrevocably loosened.

What green socialism must do, then, is to try to work *with* this grain to establish the perceived benefits of the Greek polis. This suggests decentralisation *plus* overall planning and coordination to a higher degree than centripetal market capitalism normally achieves. Socialists and anarchists do not think this intrinsically bad, provided it is open:

> Discussion of the allocation of resources in communes, collectives, councils or whatever forms the new society adopts, would not equal government. It is cooperation, surely, and to be welcomed.
>
> (Simcock 1991)

But to achieve a globally coordinated egalitarian production and distribution of goods and resources, with utmost ecological care, peace and social justice – to do this anarchistically on the basis of loose, spontaneous, direct democracy (even majority, let alone consensual) among millions of substantially autonomous communes, coops, city regions and bioregions – this stretches credibility (see below). One is therefore driven back to the position of socialists like Ryle, Frankel or Miliband: advocating some kind of state or state-like institution, albeit a Rousseauesque benign state, and one which substantially operates locally, as an enabler of local communities.

There are all too many risks in this, as history and anarchism remind us. But curtailing some 'freedoms' – even cherished liberal–anarchist ones – is indeed the price of socialism, including green socialism. To deny this is dishonest and utopian.

Against utopianism

Marxism advises us to guard against utopianism, especially in advocating unrealistic (historically blind) means of reaching our goals. Green anarchism is sometimes utopian, with its educated workforce in small communities, overcoming individualism, asserting holism and replacing competition by cooperation (a sort of red herring anyway, given already that capitalism is built on prodigious cooperation). Green anarchism would bypass the state, providing 'A realistic route out of the dysfunctional society which is the legacy of social democracy' (Atkinson 1991, 184). Marxism holds that such is not a realistic route, and history bears this out.

Neither is the postmodern 'radical relativism' which Atkinson (p. 58) counsels as the basis of green anarchist epistemology and morality. It perceives a variety of moral perspectives, *each* throwing some valid light on a problem. It, like Atkinson, replaces ethics with aesthetics and hedonism as the test of what is 'right'. It 'tells us not only to accept but even to revel in the fragmentations and the cacophony of voices through which the dilemmas of the modern world are understood'. This leads to 'the point where nothing remains of any basis for reasoned action' and 'beyond the point where any coherent politics are left' (Harvey 1990, 116). Its utopia is the ecologically-impeccable but socially-anything-goes community of child abusers which this brand of ecologism might ultimately, by logical extension, allow (Chapter 3.8).

Imagining this to be the route to that social justice which anarchists and socialists agree is prerequisite for ecological soundness really is utopian. For

A pseudo toleration of all ideas, therapies, pedagogical principles, child raising practices, cultural messages, legal statutes, etc., is both dangerous and naive.

(Frankel 1987, 192)

As Harvey shows, it leads to, not from, the conservatism and the entrepreneurial culture of neo- (not post-) Fordism. The aestheticisation of politics bases them in place (nationalism) and person (mythologised figures) rather than community, and points towards barbarous irrationalism, refusing argument and rationality and producing the death camps (Harvey 1990, 210, 273).

Some green activists sense this. Greens have lumped together all other politicians and tendencies as 'grey', says Gotts (1992). But 'the revival of Nazism exposes the unreality of this position'. And indeed it is quite possible to conceive of a right-wing politics based on environmental concerns. However, greens should recognise that their *real* allies are on the left, and put their energies into 'a living alternative to the dominance of big business and the market, rather than pretending to be above the battle'.

But creating this alternative means inverting Atkinson's (1991, 207) eco-anarchist analysis that

The heart of the problem is not, as the Marxist legacy would lead us to believe, in the organisation of production at all, but rather, as is implied in postmodern discourse, in the organisation of lifestyles and, by way of infrastructure, the creation of new concepts of community and the spread of lived practice that realises these concepts.

Atkinson seems to be talking in these last lines about *production* anyway. But he means to reject the idea of reorganising production by taking over the existing economy. So his is a green 'revolution' which, as Elliot (1983) puts it, eschews confrontation with capital – a utopian position indeed, that historically has produced no revolution. Eco-socialism rejects this postmodern way which ignores what goes on beneath the surface, instead seeking to go to the roots of power. Eco-socialism is, then, truly 'deep ecology'.

It counsels all greens to abandon 'utopian musings about alternative economic systems at the local level, communalism, small-scale alternative businesses and other counterculturalisms, and . . . recycled Schumacherism [itself recycled Mumford and Kohr]'. Instead they must now address the enormous problems of creating a socialist economics that plans

> programmes for economies of scale which confront unemployment, scarcity and an awesomely complex international division of labour, while at the same time facing the unpleasant reality of material coercion and the discipline of the wage system as the driving force realising labour power. . . .

> (Hall 1991)

In other words they should recognise the historically-contingent nature of any immediately-possible social change, and that at present 'Any state policy that relies on utopian assumptions about mutual aid and volunteerism is a formula for economic catastrophe, a descent into chaos. . . .'

Being realistic: the problem of the division of labour

In contrast to green utopianism, green socialism would revive the syndicalist and guild socialist tradition, working particularly through trades unions, Lucas-style campaigns (see below) and labour-based organisations such as the Socialist Environment and Resources Association, to confront the power of capital. It would seek socialism and ecological well being as *structural* features of an *advanced* economy.

This will indeed pose enormous problems, and Hall's point about the awesomely complex division of labour is well made. It is re-emphasised in Sayer and Walker's (1992) description of the degree to which capitalism has advanced that complexity in recent decades. And Sayer (1992) cautions us that this implies re-evaluating – though not abandoning – Marxism's abstract political economy theory. That theory, he suggests, was too idealistic (in both senses of the term). The idea of millions of 'associated producers' collectively controlling an advanced economy through advance planning without any recourse to any market regulations through prices is a 'quaint pipe dream'. That such ideas are even entertained, he thinks, illustrates a common tendency on the left to underestimate the complexity of advanced economics.

That complexity results from extreme specialisation, fragmentation, interdependence and internationalisation. It is now intractable because there are *too many* producers/consumers to be able to reconcile, in advance, supply with demand – even given the most advanced computer technology (this flatly contradicts Buick and Crump's view, as was pointed out in Chapter 3.7). Furthermore, activities cannot fully be combined and coordinated because of

229

their technological incompatibility. And motivation of producers in an advanced, complex economy will continue to be low because of the continuing anonymity of the consumers, even after socialism is established. Sayer sees little prospective diminution in that variety of interest groups which stems from the international division of labour and which leads to inequalities, rivalries and tension, and facilitates racism and gender and class exploitation.

Thus the ever deeper and more extensive division of labour – which results from that development of the forces of production which Marxism considers necessary for socialism – leads to a structure of social relations possessing causal power and abilities of its own. This structure can, thinks Sayer, *outlast* the overthrow of specifically capitalist social relations of production. He foresees that competition and externalisation of social and environmental costs could continue to be a feature of any socialist economy where there is common ownership of the means of production without concomitant control over the division of labour. Yet the latter may continue to be elusive because of uncertainty, information overload on people, free-rider problems, division of knowledge, lack of common interests and the complexity of what has to be controlled. Such problems might be eliminated

> Among small groups of self-selected people with similar interests . . . [where planned] collective control can be highly liberating. [But] when there are vast numbers of people with different interests, knowledge and material circumstances, the pursuit of collective *ex ante* control carries with it the threat of coercion and subordination to the plan.
> (Sayer 1992, 357)

And there is *no* doubt in Sayer and Walker's minds that

> Central planning is despotic and suppresses decentralised horizontal feedback between fragmented producers and users.
> (Sayer and Walker 1992, 269).

While this book does not question the realism of Sayer and Walker's analysis, it has raised doubts about this assumption that merely because centralised planning has sometimes been despotic in the past any form of it must necessarily always be so. And it has drawn attention to well-made reservations (e.g. by Frankel and Ryle) about the unfeasibility of an advanced economy being largely decentralised.

This is just one of the big problems for eco-socialism to resolve. It must also work out how the present economy can be radically transformed to socialism, without falling into the trap of utopian idealism that has so often beset green theorists. As Cole *et al.* (1983, 245–7) put it, the problem may need to be resolved somewhere between syndicalism (which perhaps overstresses control at the local workplace and therefore could not cope with the larger economic questions now addressed by finance, labour, commodity and international markets) and 'Bolshevism' (which, in Russia, overstressed dictatorial state control).

Sayer and Walker, like Frankel, consider that it needs to be resolved through a mixture of planning and markets. Notwithstanding the acknowledged inefficiency of markets when it comes to tailoring supply to genuine need, they see no prospect of a totally planned economy being anything other than inefficient as well as despotic. So a version of the mixed economy which already characterises most contemporary Western societies is their inevitable conclusion. However, while it would create profits, this does not, they suggest, have to involve profit *taking*. Profits and losses help to motivate people and regulate supply to demand. But (p. 266)

> profit-takers need not convert their profits into capital: this is a crucial distinction. Worker-owners need not use their profits to employ wage labourers and hence become capitalists; consequently production for profit need not be capitalist.

There is also a tendency, they maintain, for the left to underestimate the benefits of competition:

> Although competition doesn't always guarantee increased consumer choice and greater efficiency and more innovation, there are many cases where it does just that.

A coherent eco-socialism must acknowledge the importance of such reservations – along with all those of the neo-Marxist school detailed in Chapter 3 – about the conclusions which a crude, 'vulgar' application of Marxism could lead to. Yet Sayer and Walker's re-evaluation of Marxism might raise as many problems for ecology and socialism as it solves. And it is a little redolent of a slide towards the sort of accommodation with liberal politics and economics that ends up being largely 'revisionist' and technocentric. If socialism allowed itself to make such accommodations, then it would have little appeal to the radicalism which is rightly inherent in green thinking and is struggling to burst out. Instead, it would really do what some green critics already say that it does – that is it would merely push us back towards an outdated post-war consensus politics, economics and society: one that inherently cannot overcome the increasingly untenable ecological contradictions of capitalism.

So the radical stance on both analysis and prescriptions which comes from Marxist–socialist orthodoxy should not be so re-evaluated that the Marxist baby is thrown out with the bathwater.

Coda: reassurances about Marxism

To underscore this point, it is worth recalling that a judicious degree of Marxist orthodoxy does *not* have to lead us to that inhumane, totalitarian, inefficient nightmare which bourgeois misrepresentations of Marxism would have us believe. We should remember that there are many myths about Marxism, as Cole *et al.* (pp. 227–8) note. We can gain some reassurance by considering the more relevant ones from their list and the ripostes which they offer. They are as follows.

There is no demand in Marx's theory - only production. There is no money in Marx. Demand is in fact an integral part of Marx's theory, and there is money in it too. Marx distinguishes between money and money capital, and writes extensively about both.

Marx said that labour is the source of all wealth. In reality he criticised those who advanced this argument, and said, in *Capital II*:

> Labour is *not the source* of all wealth. . . . *Nature* is just as much the source of use values [and it is surely of such that material wealth consists!] as labour. . . .

He did argue that labour is the source of *value* as realised through exchange, as opposed to use value.

Marx, wrongly, forecast the immiseration of the working class. He discussed their *relative* material immiseration in relation to capital, maintaining that the proportion of total social wealth controlled by capital would continue to rise – as is in fact the case. And part of his discussion was about the *cultural*, not simply material, immiseration of the working class (a fact that much concerns greens also).

Marx, wrongly, forecast that the rate of profit would fall continuously. In fact he discussed the *tendency* for the rate of profit to fall.

The abstract labour theory predicts that capitalism will break down and it hasn't. The theory does not say this, but that capitalism is prone to crisis, that the resolution of one crisis leads to another, and that there is a fundamental contradiction in capitalism between the social nature of production and the private nature of appropriation.

This last should perhaps be the central point which eco-socialism must labour to communicate to the world – with a crispness and clarity that has so far been lacking from the green critique.

5.3 ECO-SOCIALISM SUMMARISED

From the above, and Chapters 3 and 4, we can summarise major principles of eco-socialism: to be recommended to all radical greens, mainstream and ecoanarchist.

Eco-socialism is anthropocentric (though not in the capitalist–technocentric sense) and humanist. It rejects the bioethic and nature mystification, and any anti-humanism that these may spawn, though it does attach importance to human spirituality and the need for this to be satisfied partly by non-material interaction with the rest of nature. But humans are not a pollutant, neither are they 'guilty' of hubris, greed, aggression, over-competitiveness or other savageries. If they behave thus, it is not by virtue of unchangeable genetic inheritance, or corruption as in original sin: the prevailing socio-economic system is the more likely cause. Humans are not like other animals, but neither is non-human nature external to society. The nature that we perceive is *socially* perceived and produced. Also, what humans do is natural.

Thus alienation from nature is separation from part of *ourselves*. It can be overcome by reappropriating collective control over our relationship with nature, via common ownership of the means of production: for production is at the centre of our relationship with nature even if it is not the whole of that relationship. We should not dominate or exploit nature in the sense of trying to transcend natural limits and laws, but we should collectively 'dominate' (i.e. plan and control) our relationship with it, for collective good.

The eco-socialist response to resource questions is not merely to fix on distribution, as commentators like Eckersley suggest. It says that there are no ahistorical limits of immediate significance to human growth as *socialist* development. But there are ultimate natural constraints which form the boundaries of human transformational power. Additionally, each form of social–economic organisation has its specific way and dynamic of relating to its own specific set of historical conditions, including the non-human environment. So the natural limits on a given mode of production are not universal limits, of a universally similar kind, on all modes of production. Changing the mode of production means changing many needs, and therefore the resources to fill them, and also the set of ecological problems which must be solved. Eco-socialism would change needs, redefining wealth along William Morris's diverse lines, which also include a 'bottom line' of reasonable material wellbeing to all. But all these material needs can be met through socialist production, because there are limits to them, although generally human needs will always become more sophisticated and richer in socialist development.

Production and industry are not to be rejected *per se*. If unalienated, they are liberating. Capitalism initially developed productive forces, but now it hinders their unalienated and rational development. It therefore must be replaced by socialist development where technology (a) is adaptive to all nature (including human) and not destructive of it, (b) strengthens the competence and controlling power of the producers.

Planning is vital in socialist development, through an enabling 'state' or similar institution:

> stateless, moneyless, small-scale communes or other informal alternatives are not viable without the complex administrative and social structures necessary to guarantee democratic participation, civil rights and egalitarian coordination of economic resources. . . .
>
> (Frankel 1987, 270)

This last requirement will involve some world-wide exchange and reciprocity, for the resource development and distribution according to need, not profit, which eco-socialism demands. Production will not be built on wage slavery, but on volunteered labour, which most people will want to give to fulfil themselves and relate to others. Hence individual desires will largely be reconcilable with the strong community ethos, though some present 'freedoms' such as that to own land will be lost, and people may feel peer pressure not to be free riders.

Eco-socialism defines 'the environment' and environmental issues widely, to include the concerns of most people. They are urban based so their environmental problems include street violence, vehicle pollution and accidents, inner-city decay, lack of social services, loss of community and access to countryside, health and safety at work and, most important, unemployment and poverty. These problems are not all specific to capitalism, but are more crippling and globally widespread in it than in past modes of production. Therefore the basic socialist principles – egalitarianism, eliminating capitalism and poverty, resource distribution according to need and democratic control of our lives and communities – are also basic environmental principles. Part of the definition of true communism is that people will not experience an environmental crisis through it: non-human nature will be changed but not destroyed, and more pleasing environments will be created than destroyed.

Eco-socialist strategies for achieving communism may vary. But they have in common that they appreciate the potential need to confront rather than try to bypass capitalism: and that we have great power, communally as producers, to build the society we want. Hence the labour movement must be a key force in social change, rediscovering its potential in this respect, and resurrecting its character as an *environmental* movement, historically demonstrated in unionism, utopian socialism and the back-to-the-land movements, for instance.

The approach to social change and historical development, while not discounting the power of socialisation, education and ideas, will be materialist – recognising the key importance of economic organisation and material events in influencing consciousness and behaviour. Latent class conflict is still potentially a powerful force for change, and class analysis is still important, albeit applied from a global perspective.

Trying to smash capitalism violently will probably not work while capitalists control the state, so the state must be taken and liberated in some way for the service of all. There are limits to achieving this by attempting a revolution in mass consciousness via education and exemplary lifestylism. Neither can involvement in managing capitalism produce fundamental solutions to environmental crises. Nor will a dictatorship of the proletariat, initiated by a vanguard which then becomes the dictator, be acceptable.

An ecologically sound socialist society will not come until most people want it enough to be prepared to create and maintain it. Probably, and regrettably, the biggest catalyst will be the failure of capitalism (a) to produce 'the goods' which it promises, for even a small minority (b) to create a physical and non-material environment for the rest which is tolerable enough to contain discontent. But the development and extension, now, of an oppositional eco-socialist line of ideas and actions will help the change and help to reduce the future casualties of capitalist regimes.

5.4 ECO-SOCIALISM IN PRACTICE

Books which analyse our environmental predicament are on a hiding to nothing. They can be criticised either for not suggesting a coherent and feasible action programme, or, if they do have a programme, for being naive and/or anodyne about what could and should be done. This is partly because liberal–capitalist assumptions about the purpose of life and how to live it have gained such hegemony that any attempt to move towards a society based on alternative assumptions does seem either undesirable or futile. This is inevitable in any culture – even most people in Orwell's *1984* are 'happy' with their lives and/or see them as unalterable. Conversely if action programmes do sound practicable and achievable, then they probably subscribe implicitly to the tenets of the existing socio-economic–political paradigm, and are unlikely to threaten its continuance.

This book is about eco-socialist theory; however, part of that theory is to point out how radical social change is not only possible, but has consistently happened and is always likely to occur. Also, the important purpose of the Marxist theory on which much eco-socialism is predicated is to foster 'praxis': the universal action by which humans shape their history and themselves. So even a book on eco-socialist theory must at least suggest the kinds of environmental action which its analysis favours.

Eco-socialists might reasonably support most environmental actions designed to change economics, politics and society: if only because to do something is usually preferable to inaction. But at the same time they will be on guard against actions which ultimately might reinforce the status quo directly ('ecotage' may well do this), or indirectly by encouraging a false consciousness that radical change is being effected when it is not (green consumerism, 'Band Aid'-style charity and the like).

But from the analysis it follows that the most potentially fruitful kinds of action are those which emphasise people's collective power as producers, which directly involve local communities (particularly urban) and increase democracy, which enlist the labour movement and which are aimed particularly at economic life. The examples outlined below each do some of these things. None are totally satisfactory and theoretically unimpeachable, but they all deserve eco-socialist support and emulation. Some further examples may be found in Ekins's (1992) collection of descriptions of 'new' approaches to development. Many are initiatives by small, local communities and cooperatives and they represent a constructive response to the disastrous social and environmental effects of capitalism, though whether, collectively, they add up to the 'New World Order' of his title is doubtful. They are, as he says, a way of acting *within* an economy, rather than of changing it. But to be more positive and dialectical, perhaps they represent part of that order whom the existing economic and social arrangements do not satisfy and which will eventually, by struggle with the existing order, produce a new socialist synthesis.

Alternative production, trades unions and the community

Trades unions and the labour movement can play both negative and positive roles in environmental campaigns. Even the emasculated union movement of 1980s' Britain has said 'no' to some developments, notably sea dumping of nuclear waste and landing and incinerating foreign toxic waste on British shores. But probably the most effective 'green bans' took place in Sydney, Australia in the early 1970s. Then, many of the city's residential areas were threatened by high-rise speculative commercial development, often using outside capital and backed, or initiated, by local government.

The first ban happened after local women residents from the Hunter's Hill suburb approached an official of the New South Wales branch of the Builders' Labourers' Federation, telling him of their concern over threats to develop a bank of the Parramatta River where stood a last remaining area of bush vegetation. They had been encouraged by reading a statement from the BLF which asserted that as workers who had raised the buildings, they had a right to express an opinion on social questions related to the building industry (Mundey 1981). This meant that the union interpreted broadly its brief to 'improve members' conditions', appreciating that this involved better living environments. In this first environmental ban labourers refused to work on sites where local communities would suffer substantial degradation of their environment from the proposed development. Others followed, in the Eastlakes, The Rocks and Wooloomooloo districts. At The Rocks the opposition was the State Government-owned Sydney Cove Redevelopment Agency. The Government refused to consider alternative, low-cost, medium-density housing. The ban was backed up by residents refusing to be evicted, and when scab labour was used there was a joint union–residents' occupation of the old Playfair building to stop its demolition.

The character of many areas of the city was maintained as a result of the successes of these bans in getting agreement to more environmentally-sympathetic developments, often involving rehabilitation. The bans spread, to Newcastle, Hobart, Perth, Fremantle and Adelaide. So did publicity about them, and in 1976 the old Birmingham Post Office building in England was saved from destruction by a coalition of building workers and environmentalists.

However, in 1974 the employers sought deregulation of the BLF and the Sydney branch leaders were victimised. The establishment had realised the serious potential threat to their interests: 'What's now coming is a new type of personal property in which you have some "rights" just living in a place', said Sydney's lord mayor disdainfully (Bolton 1981). Absolutely so – and about time! But what really ended the bans was the end of workers' relative solidarity brought on by economic downturn: unemployment, it seems, invariably wrecks worker solidarity just when it is most needed.

Several ingredients made the bans successful for a while according to Mundey. Most vital was the rank-and-file interest in the union and what it was doing. And members were prepared to strike over their (broadly interpreted) conditions, and to back each other up – when one site was targeted all other sites belonging to that developer were also hit. Unfortunately there was no similar inter-union solidarity. Health and safety, of course, was particularly interesting to building workers, because there were so many deaths and injuries in the industry. But the union had a history of involvement in broader social issues, having stopped work to stop the Vietnam war in the 1960s, and having supported aboriginal land rights and homosexual rights in the 1970s. The union's social conscience embraced a desire to help other citizens, not just members, and a willingness not to be blinkered by pure monetary considerations. The bans' effectiveness was enhanced by the involvement of middle-class people. There would be no ban unless a majority at a residents' meeting had asked the union for it. Even the National Trust gave covert support. And there was the communist origin of many of the leaders. This gave them a penchant for democratic procedures and actions – they cut their own pay to be on a par with the members, who had voted them in and the established union leaders out.

Mundey (p. 148) believes that the middle classes are naive about where real power resides, but that most unions are not yet sufficiently aware of the seriousness of the environmental crisis, to which their work might contribute. He therefore advocates a 'red–green' coalition of the two, to eliminate each of these blind spots.

But the trades unions' blind spot may be deep seated, in their concern over a perceived 'threat' to jobs posed by environmentalism. This perception goes wider than unions: 'Before discussing the threat to the ozone layer due to the First World's faults', said the mayor of a Rio de Janeiro slum suburb of heavy industry and unbelievable poverty and degradation during the 1992 Earth Summit, 'I would like to know how to give food and dignified living conditions here' (Etchart 1992). Small wonder that guests at this event were greeted by billboards saying: 'Ecologists go home'.

Eco-socialism must meet concerns over jobs, not just by offering the prospect of informal work, and some environmentalists have tried to do so. 'Environmentalists for Full Employment' operated from 1975 to 1984 in America, on the premiss that 'It is possible simultaneously to create jobs and conserve energy and natural resources' by putting 'an end to the exploitation that environmentalists and unions have, heretofore, fought independently' (Grossman 1985, 86). It had some success in mollifying the anti-environmentalist stance of the US's major unions, particularly over the nuclear issue (which has greatly split unionists and labour in Britain). The key to this was, says Grossman, not a leadership-dominated but a democratic coalition, and through working locally in defence of union causes and for more job-intensive alternative products and production methods.

Of course the most well-known effort to convert industry from employment in environmental destructiveness was the Lucas Aerospace Combine Shop Stewards Committee's plan for alternative production. As Frankel (1987, 49) says, this has become an uncritically accepted sacred cow of the left, like Mondragon (see below). Frankel argues that it is not applicable to the majority of private and state workers, who do not directly produce commodities. Nonetheless it has principles which are, so it is worth noting these from Wainwright and Elliot's (1982) account, and just how compatible the plan was with ecologism.

The plan was widely publicised in 1976, but it evolved before and after then. It documented the productive resources at Lucas Aerospace, a firm making hardware for the 'defence' industry, which was therefore subject to the vagaries of government policy as well as to the market, and had declared one of its periodic redundancy programmes. It analysed the problems and needs faced by Lucas workers in the face of this government-market context. It assessed the general social needs which could be met by Lucas's resources and it detailed products, production processes and the employment development pro-grammes which could meet those needs.

Many products were suggested (Wainwright and Elliot, pp. 100–7); some directly quantifiable as ecologically benign alternative technology, for instance solar power and wind generators, heat exchangers, a road–rail vehicle, a hybrid petrol–electric car and an airship using jump jets to avoid helium waste. More important were the proposed production processes, which (a) would not waste energy and raw materials; (b) could be labour intensive, to avoid structural unemployment; (c) would be organised in non-hierarchical and non-alienating ways; (d) would involve discussions with those for whom the products were intended; (e) would break down distinctions between scientific and manual and skilled and non-skilled jobs; (f) would develop the skills and fulfilment of producers. This was all very green – and socialist.

The Lucas shop stewards emphasised that the market did not produce socially useful things which were not profitable. Indeed, a survey (Shenfield 1971) had concluded that out of twenty-five large British companies none

> had any doubts that their primary objective was to be efficient and profitable and that being socially responsible would serve no useful purpose if it hindered these overall goals.
>
> (cited in Wainwright and Elliot, p. 241)

This is salutary for all would-be green consumerists/capitalists, suggesting the limits of their reformism, as cheerfully conceded by one of their gurus, Richard Adams: 'If you can't sell it there's not much point in making it' (Hoult 1991, 44).

The Lucas plan was rejected by all elements of the British establishment, including management, conservative trade unionists (one *complaining* that it would change British society!), academic 'leaders' and most of the Labour Government, whose initial encouragement was sustained by only a few, like Tony Benn. This is unsurprising since it was revolutionary, proposing industrial restructuring in the interests of labour; redefining wealth by rediscovering William Morris's definition ('working cheerfully at producing the things we all genuinely want'); redefining, therefore, economic rationality; challenging labour vanguardist views that average workers can do little more than to describe their grievances; reasserting working people's right to associate (across unions); exposing the hidden values behind seemingly neutral, technical, 'rational' management and challenging its right to manage, at least without accountability to workers: as Lucas workers at Shipley are reported to have said:

> In our experience management is not a skill or craft. It is a command relationship, a habit picked up at public school, in the church or from the army. And we can well do without it.

> (Wainwright and Elliot, p. 89)

Through such principles, and in its combination with environmental groups, the Lucas people laid down a true eco-socialism, pointing to

> another form of planning and economic decision making, very close to Marx's [*Capital*, vol. 3] notion of 'associated producers rationally regulating their interchange with nature, bringing production under their common control instead of being ruled by it as by the blind forces of nature' [cited in Wainwright and Elliot, p. 254, who also ask, p. 238]: Could we really expect technical elites nurtured on a diet of weapon systems development, a criterion framework of cost efficiency and a free enterprise management ethos really to address themselves to the technical tasks involved in providing human dignity and a peaceful planet?

The answer is: of course not – and Marxist analysis ensures that we remember this even though so many greens forget it. Marxism emphasises the social context of science and technology, showing, as in the Lucas case, that usually it is not technologies (e.g. district heating schemes, the heat pump) that are radical, but their organisational context, which may be. Different technologies of themselves will not change society (even though they may lend themselves more or less to socialist or capitalist productive relations), unless, as in the Lucas case, workers specify and control different uses and productive methods. A windmill produced by and for a giant private electricity company is not an 'alternative' technology.

As Frankel suggests, some features of the Lucas plan were uniquely applicable to that industry, with its skilled workforce producing versatile

239

technology in small batches in an industry not fully exposed to the market. But it does illustrate how more widely relevant workers' control could become if the inter-union collaboration that was a feature were to be extended. Developing conditions like structural unemployment might also be thought significant, although the acceleration of this trend in the 1980s brought contraction rather than expansion of the alternative production movement. The importance of left-thinking trade union and shopfloor radicals needs to be acknowledged, as with the green bans. They were the progressive, inventive people, who wanted their own jobs (in 'defence') to be cut, and to be allowed to replace them by their own efforts. And they were doing it for traditional trades union reasons, to improve wages, conditions and prospects of the workforce – battling against capital's footloose search to evade the potential power of organised labour.

The Lucas plan stimulated many other developments, including alternative production plans in other branches of the 'defence' industry, which became a limp part of the Labour Party's 1992 programme – far too late to avoid the unemployment consequent on the cold war's end.

There was a Tyne–Wear plan for a combined heat and power station. Here a group of workers in the early 1980s, combined with local tenants' groups, sought to build a positive alternative, and environmentally sound, strategy to combat structural unemployment by 'exerting their collective autonomy against the logic of capital' (Byrne 1985). And there was a Centre for Alternative Industrial and Technical Systems founded at North East London Polytechnic, one of the more open-minded academic institutions in this instance, in 1978. Eventually such initiatives became transmogrified into Greater London Council (GLC) schemes (like the Greater London Enterprise Board) to assist innovation and provide products to meet social need, and other similar initiatives in Sheffield and Coventry. They mostly faded with the passing of the GLC and other leftist local authorities, since they were an aspect of a fitful municipal socialist movement which irritated the established order, and against which the British Conservative Government has consistently battled for two decades.

Municipal socialism

Socialism through the town hall is a contradiction in terms to some socialists and anarchists, and we have noted why. Nonetheless, many left-oriented Labour local councils were elected in the 1980s, and they provided something of an antidote to the Conservative programme of centralisation of political and economic power and structural violence against the poor and weak. Many of their policies aimed, among other things, at two objectives compatible with eco-socialism. One was improving the urban environment – 'natural' and built, social and economic. The other was democratically to involve and enable local

240

communities to improve their cities. This was often done in conjunction with environmental groups.

The Labour-dominated Association of Metropolitan Authorities (AMA 1985) declared that its policy was to encourage different local government departments to join together in order to upgrade housing, renovate parks, develop derelict land, create play centres and improve access and rights of way in countryside on the urban fringe. In this it wanted to promote self-management by local and community groups and clubs.

AMA gave many examples of this work: city greening in Wakefield and Leeds; Bradford's 'operation eyesore', landscaping derelict inner-city areas; water meadows and leisure centres, and country parks and nature reserves in Barnsley and London; estate refurbishment in Bootle; sports grounds, horticultural and arboricultural sites in Tameside, Lancashire; reclamation and tip restoration, and nine million trees planted in ten years in Manchester; linear canalside walkways in Birmingham; and financial support for voluntary nature conservation bodies in Tyne and Wear.

However, the size of AMA's task in some areas is enormous, with central government continually emasculating the authorities, politically and financially. The City of Bradford (1989) is but one example of a council attempting to 'Go Local' to people in their own 'community environmental projects' and 'neighbourhood forums'. But it faces hostility from the right, as well as some apathy from the people whom such involvement is intended to benefit. Despite glossy free 'green' newspapers despatched to every household, rubbish is more likely to find its way onto the street than into a recycling bin. Bradford's people have more pressing preoccupations, created by economic oppression. However, other councils like Sheffield and Oxford, have made more progress with recycling, if not with eliminating vehicle pollution.

Municipal socialism is not revolutionary, but it does embrace some important eco-socialist principles. Its definition of the environment centres on where people live, rather than the 'natural', it advocates a democratic socialist perspective, to 'regulate, control, supplement or eliminate markets', it tries to affect local democracy and respond to that democratic wish, it is involved in distributional aspects of wealth, and questions of jobs and the environment, and it seeks to change the pattern of land ownership (Smith 1987). Neither is its environmental programme antipathetic to the labour movement, for it creates jobs, in projects like the London Energy and Employment Network, the Lancashire Enterprises proposal to regenerate the Leeds–Liverpool Canal corridor, and the Sheffield Centre for Product Development and Technology Resources.

Alternative social and economic arrangements

All of the above meet Elliot's eco-socialist criterion of 'confronting capitalism' in some way. But the alternative, more anarchist/utopian socialist strategies of

241

bypassing or prefiguring, can also deserve eco-socialist support, particularly since many are aimed at collective economic life.

Another of the left's 'sacred cows' in this respect is the Mondragon complex of cooperatives in Spain. This involves a federation of over 100 worker-cooperatives, employing about 20,000 people. They produce many goods, notably domestic appliances, which compete successfully in capitalist markets. Inspired by utopian socialist experiments, they originally embraced many socialist principles. Directly mandated workers' councils hired and fired managers, wage differentials were very small, capital could not be taken out of the system, 'wages' were a given proportion of any financial surpluses generated, much of the remainder of the surpluses went to fulfilling the main purpose of the system – to create more enterprises and jobs, finances were facilitated by a vital secondary cooperative, a 'people's bank', and a whole series of secondary coops provided, internally, retail, educational, health and welfare services. Additionally, when coops grew beyond a certain size they had to divide (Campbell *et al.*, 1977). However, Mondragon, like so many similar, if less ambitious schemes, has experienced the phenomenon of 'goal displacement'. From an attempt to rehabilitate the devastated Basque community after the Spanish civil war has now come a set of businesses dedicated to survival on the basis of capitalist economic premisses. The wage differentials have eroded from 1:3 to 1:6–7, for example. There is insufficient reinvestment of profits in the social structure, while the further education institution now just teaches technical skills and not the value of cooperation (Encel 1990).

Indeed, Ward (1991) thinks he sees in Mondragon the same goal displacement that in Britain changed the 'Cooperative Permanent Building Society' into the fully commercial profit-oriented 'Nationwide' society, and asks

> Do we judge that success is the kiss of death for any cooperative enterprise, and that failure is glorious, or do we need an entirely new set of criteria?

It is true that, theoretically, coops are part of a socialist programme. They can give people control over their own job, and therefore job satisfaction and better working conditions. They can release untapped potential on the shopfloor, creating greater efficiency and better industrial relations. And they can be devoted to socially useful production and the conservation of health and environment. But according to the Open University (1986) the performance in Britain has provided 'little evidence so far that coops are the platform for socialist ownership at national level'. This is partly due to insufficient support from the labour movement. But it also stems from the fact that many coops survive through self-exploitation rather than efficiency or good cooperative management, which is difficult and complex. Furthermore, though many coops produce and sell 'alternative' things (e.g. in wholefood shops and restaurants), they have not spread an alternative ideology of socially useful production (far from it in some cases). They thrive because they have learned skills and values of conventional business, which is unsurprising because many have been formed purely 'defensively' in response to recession in the conventional economy, not positively for political reasons.

Frankel (1987, 31) encapsulates the problem:

> All enterprises governed by market mechanisms (whether cooperatives, or publicly or privately owned) are pressured into competition and constant growth in order to sustain income, market share and hence survival. Only a planned economy can avoid the problems of overproduction, labour shedding, pursuit of international markets and crises in profitability which market mechanisms produce. . . . It is extremely difficult to imagine any self-sufficient socialist society operating according to simple barter or non-market exchanges between cooperatives.

He believes that schemes which propose doing without both state planning and market mechanisms are utopian. Perhaps the same may be thought of those which advocate a no-money economy, or one without some universal unit of work equivalence. Without judging on this question, it is worth pointing out that there is a potential 'halfway stage' between 'free enterprise' and the moneyless socialist economy. This is the local currency.

There are many examples in North America or Britain. Each member who transacts through the one-time Totnes Green Pound Bank, for instance, had a chequebook to pay for goods and services, and could charge for their labours in green pounds, so the currency stayed in the community's circulation and did not pay for national scale economics.

Indeed, local currencies prevent the inter-regional appropriation of surplus value which is at the root of so many environmental problems. To this extent they must be compatible with eco-socialism. Socialists might justifiably object that they do not prevent exploitation or accumulation within a region. However, some schemes have been devised with built-in periodic currency changes, to obviate this problem.

Button (1988, 255–6) summarises local currency characteristics:

> 1 The agency maintains a system of accounts in a quasi-currency, the unit being related to the prevalent legal tender; 2 Member accounts start at zero, no money is deposited or issued; 3 The Network agency acts only on the authority of a member in making a credit transfer from that member's account into another's; 4 There is never any obligation to trade; 5 A member may know the balance and turnover of another; 6 No interest is charged or paid on balances; 7 Administrative costs are recovered in internal currency from members' accounts on a cost-of-service basis.

Given its concern about the removal of economic and therefore political power from old core industrial areas, local currencies tied to local cooperative

243

networks might deserve more support from trades unions and the labour movement.

5.5 POSTSCRIPT: THE WAY FORWARD FOR RED-GREEN POLITICS

Since the rise of mass environmentalism in the 1970s there has been a tension between red and green, which has produced both conflict and attempts at reconciliation and collaboration. Thus red–green alliances and networks have abounded. They are still enthusiastically promoted. Cook (1992), for instance, writes of the 'uncanny' similarities between the two movements, both with fundamentalist and realist wings. He declares that

> the delta formed by the confluence of these two tributaries [greens and socialists, not greens and 'democratic' labour parties] transformed by the ideas of feminism and anti-racism, is a fertile political soil that is unrepresented in British politics.

I suspect that this may be politically as well as geomorphologically wide of the mark. The principal rivers feeding into this confluence – Marxism and anarchism – do sometimes uncannily resemble each other. They travel in roughly the same direction. But to get there, they insist, requires different routes over very different terrains: so much so that they may be radically different sorts of rivers. I think that if there is to be any effective red–green coalition, the radical differences in their perspectives must be thoroughly aired, and not glossed over.

The red river assumes limits to human needs and that they all can be met. The green assumes limits to growth. Red is for a modified 'Enlightenment product' and modernism. Green is substantially postmodern. Red is absolutist – for socialist development. Green is a mix of naturalistic absolutism and social relativism. Red's view of nature and society is monist. The green view professes monism but practises dualism. And so on. Quite a lot of these differences mirror the differences between Marxism and anarchism and utopian socialism.

Such a bald statement, however, ignores the fact that both anarchism and Marxism are broad churches. If you accept Woodcock's classification of the former (Chapter 4.1) – not all anarchists do – then anarcho-communism and -syndicalism are the forms most compatible with Marxist socialism. However, the former does not constitute an acceptable compromise for 'orthodox' Marxists at least, for it spells out a commonly acceptable goal but unacceptable means of attaining it. Syndicalism is perhaps more acceptable because it implies methods centred around collectivity and production. Here, I think, there is very much common ground, ripe for exploration. Even (especially) in the slump of the early 1990s, there is much more scope than red–green politics currently allows to revive the alternative production and green ban movements, for instance.

But the disagreements about fundamentals are also strongly felt on the anarchist-green side. Carter (1988) has presented them as a coherent critique of Marxism with an anarchist counter-theory of history and social change. He maintains (p. 263) that 'the problems encountered in Russia are directly linked through Lenin to Marx. It is the whole Marxist paradigm which needs to be rejected'. Marxism, he says, has an epistemological weakness, inherent in its materialism, which overemphasised labour and production as principal means of knowing and interacting with the world. This is inadequate. Furthermore, the predictive value of Marx's theories – labour theory of value, theories of class and exploitation, and of the state – is poor, because they are all special cases of more general theories. And Marx's 'scientific' theory of history predicts on the basis of supposedly universal processes. However, merely because something has happened (e.g. formation of an immiserised proletariat) does not mean that it always will.

Other anarchists, and greens, incessantly attack Marx for crude economism, historical determinism, inherent illiberalism and utopian and totalitarian tendencies, modernism, over-collectivism and so forth. I have noted such criticisms in Chapter 3 and shown that by no means are these or Carter's strictures applicable to all interpretations of Marx, 'humanist' or 'orthodox'. If this interpretation can be accepted on the green side, this would obviously create much common ground.

But it would still leave the issue of anthropocentrism, which Eckersley correctly pins on all varieties of Marxism and eco-socialism and even eco-anarchism, because they ultimately attribute instrumental rather than intrinsic value to nature. She concludes that only ecocentrism provides a green political theory comprehensive enough to solve the ecological crisis, because only ecocentrism is *eco*-centric. Hence even the early Frankfurt School of Marxism, whose norms of what is right are based on the outcome of rational communication, should be rejected. Eckersley concedes that such rationality *would* in fact provide an ecologically better society, and protection for nature and people, but the problem for her is that its moral referents are human. Hence, she claims, it would not protect species that serve no human purpose. But the difficulty with this objection is that it seems to assume that 'human purposes' are likely to be interpreted narrowly and economistically even in communism, which does not follow at all. 'Instrumentalism' could mean that we wish to protect 'useless' species because we have a mind to their future possible utility (intergenerational justice), or, as Grundmann suggests, because they give us spiritual or aesthetic satisfaction, or just because we like the idea or hold the value that it is morally right to do so. Communism as portrayed here is the economic and social system which will allow us to do all this, much more than the market-oriented economy (planned or no) which Eckersley proposes. However, such eco-socialism still is not good enough for the intrinsic value theorists, for even if it does practically produce the same ecologically benign society that these theorists seek it is morally defective for assuming that humans are more worthy than non-human nature.

Objections to the intrinsic value theory, which hinge on the fact that it is impossible for us to be *other* than anthropocentric, have been described above. To prefer to give equal moral value to non-human and human natures is still a *human* preference. Deep or 'transpersonal' ecologists like Eckersley or Fox counter that this point is trivial and tautological – it is an 'anthropocentric fallacy'. Of course, they agree, we are talking about human judgements: this does not mean that nature does not have its own interests, the development and protection of which are equally as important and just as the interests of humans. Furthermore, the anthropocentric fallacy errs by conflating the identity of the perceiving subject with what is perceived.

My response to this is that of course there is such conflation. To try to gainsay this is to advocate dualistic thinking – the subject (human)–object (nature) separation which ecocentrics ostensibly reject. And the issue is not trivial, but substantive. Even wanting not to privilege humans over nature when the chips are down is inescapably a *human* preference: we have no evidence that non-human species might perceive each other so unselfishly. So, despite Kropotkin, we can say only that it seems certain that each species privileges *itself*. Watson is right, therefore, to emphasise that proposing that humans alone should be different in this respect is a human value – so it must be justified in human terms, as better than other values. It might be argued that such a value will produce a better, richer 'nature' for us to enjoy, but this must be *argued*, not assumed, and it has yet to be demonstrated that proceeding from truly socialist premises could not, too, produce the same rich non-human nature. The socialist argument cannot be destroyed merely by referring to what went wrong behind the old iron curtain.

My own preference is to privilege humans, strongly believing that to do so is to achieve similar outcomes, materially and spiritually, to what ecocentrics want. I prefer this because the idea that Gaia continues on after humanity has destroyed itself gives me no satisfaction. And I think that privileging non-human nature seems to lead towards a slippery slope – either to middle-class nature-protection elitism or to plain misanthropy. Eckersley strenuously denies misanthropy in ecocentrism, on the logical grounds that to want justice among *all* creatures implies also wanting human social justice. Socialism, she persuasively argues, is subsumed by the more comprehensive ecocentrism.

Yet *in practice*, I feel, ecocentrics such as bioregionalists, deep ecologists or New Agers do tend towards misanthropy. Although they claim that 'of course' they want social justice, they emphasise 'nature's' interests – or the interests of indigenous peoples that they regard as closest to nature. This leads to the apoliticism, naivety and so forth, which have been outlined in Chapter 3.9. The problem in Britain, which is mirrored elsewhere, is that

> The Green Party is divorced from its potential supporters in Britain. Any fruitful dialogue there might have been a few years ago – before the environment groups became competing think-tanks – has gone. Although policy statements from the green groups and their increasingly political analyses of trade, debt and industry all move closer to virulent anti-capitalism, they remain pathologically frightened of committing themselves to British party politics.
>
> (Vidal 1992)

And by extension they remain frightened of committing themselves to anti-capitalist positions on the sort of fundamental political questions which were outlined in Chapter 1.2.

Yet the 1992 Earth Summit showed quite clearly the importance of such questions, and that social justice is not an area merely to be subsumed under the banner of justice to all creatures in an end-of-ideology thesis. Third world countries are, simply, and rightly, refusing to make short-term economic sacrifices to protect their rainforests when advanced capitalist countries refuse even to put their own consumer lifestyles on the agenda. Therefore social and redistributive justice has now become the *central* issue in achieving the kind of relationship with nature which ecocentrics want. Hence prioritising *social* justice must be the essentially common ground for all red–green alliances. And in view of the broader green movement's failure generally to effect radical change or to be widely appealing all over the world, it, too, should make this more openly its priority.

It follows, therefore, that mainstream greens and green anarchists must accept several positive things from Marxism. There are the socioanalysis of capitalism and the conception of the society–nature dialectic – both of which are powerful, insightful and accurate. Then there is the commitment to socialism. And there is the possibility of a social change meta-theory which takes relevant aspects of Marxism and at the same time develops a strategy that will avoid ecological destruction. Humanist/neo-Marxists in general, and the *Capitalism, Nature, Socialism* group particularly, are working on this project.

But sometimes the project displays potential problems. It tends to accept without much discussion the simplistic ecocentric limits to growth/overpopulation theses (which strongly resurfaced at the Rio Summit, predictably, from a British Government with a long track record of placing on the disadvantaged the responsibility for their own plight). However, the socialist contention that there are abundant resources to meet *everyone's* needs, when you remember that 'needs' are to be divorced from our present market-oriented conception of them, has not been convincingly disproved.

The red–green project is also in danger of dismissing too easily the existence of a working class, and its potential in social change: replacing it, in its historical important position, by the bourgeois new social movements. Neither

247

may it ascribe any key role to economics and materialism in the analysis of history and social change, while at the same time elevating idealism. It may even be ambiguous about capitalism's role in environmental degradation and the need to replace it and market economics in any ecotopia.

In other words, it may be abandoning much of Marxism altogether. As I have said, this would not worry me at all, if it nonetheless creates a coherent eco-socialism. However, I think that there is still much in relatively orthodox Marxism, as interpreted by the sources I have seen and cited, which is vital to eco-socialism and should not be summarily rejected. Though it does not constitute a complete eco-socialist theory of itself, to cast Marxism's perspectives on the green problematique can at the very least constantly provide an antidote to the vagueness, incoherence, woolly-mindedness and occasional vapidity that can invade mainstream and anarchist green discourse.

REFERENCES

Abrams, P. and McCulloch, A. (1976) *Communes, Sociology and Society*, Cambridge: Cambridge University Press.

Albury, D. and Schwartz, J. (1982) *Partial Progress: the Politics of Science and Technology*, London: Pluto Press.

Alexander, D. (1990) 'Bioregionalism: science or sensibility?', *Environmental Ethics*, 12, 161–73.

AMA (Association of Metropolitan Authorities) (1985) *Green Policy: a Review of Green Policy and Practice in Metropolitan Authorities prepared by the AMA Green Group*, London: AMA.

Anderson, V. (1991) *Alternative Economic Indicators*, London: Routledge.

Andrewes, T. (1991) 'Making headway', *Green Line*, 87, 4–5.

Ashton, F. (1985) *Green Dreams: Red Realities*, Milton Keynes: Open University Network for Alternative Technology and Technology Assessment.

Atkinson, A. (1991) *Principles of Political Economy*, London: Belhaven.

Atkinson, A. (1992) 'What chance bioregionalism?', *Green Line*, 100, 8–9.

Attfield, R. (1983) *The Ethics of Environmental Concern*, Oxford: Blackwell.

Bahro, R. (1986) *Building the Green Movement*, London: Heretic Books.

Baldelli, G. (1971) *Social Anarchism*, Harmondsworth: Penguin.

Baritrop, R. (1975) 'Anarchism, past and present', recorded lecture given to the Socialist Party of Great Britain, London.

Baritrop, R. (1976) 'Pollution and the environment', recorded lecture given to the Socialist Party of Great Britain, London.

Baugh, G. (1990) 'The politics of social ecology', in Clark, J. (ed.), *Renewing the Earth: the Promise of Social Ecology*, Basingstoke: Green Print, 97–106.

BBC (British Broadcasting Corporation) (1987a) 'Choices', discussion on the environment.

BBC (British Broadcasting Corporation) (1987b) 'What price progress?', narrator, Bob Geldof, London: BBC.

BBC (1992) 'The Noble Savage', Horizon, London: BBC.

Benn, S. (1982) 'Individuality, autonomy and community', in Kamenka, E. (ed.), *Community as a Social Ideal*, London: Arnold, 43–62.

Benton, T. (1989) 'Marxism and natural limits: an ecological critique and reconstruction', *New Left Review*, 178, 51–87.

Bermant, C. (1991) 'Green zealots are my deadliest foes', *Observer*, 6 January, 15.

Bernari, C. (1990) 'Worker worship', *The Raven*, 3(3), 213–23.

Biehl, J. (1991) 'Farewell to the German Greens', *Green Line*, 89, 12–13.

Blackstone, T., Cornford, J., Hewitt, P. and Miliband, D. (1992) 'The economics of harmony', *Guardian*, 18 February, 17.

Blowers, A. (1984) *Something in the Air: Corporate Power and the Environment*, London: Harper and Row.

Blowers, A. and Lowry, D. (1987) 'Out of sight: out of mind: the politics of nuclear waste in the UK', in Blowers, A. and Pepper, D. (eds), *Nuclear Power in Crisis*, London: Croom Helm, 129–63.

Bolton, G. (1981) *Spoils and Spoilers: Australians make their Environment 1788–1980*, Sydney: Allen and Unwin.

Bonnett, A. (1989) 'Situationism, geography and poststructuralism', *Environment and Planning D: Society and Space*, 7, 131–46.

Bookchin, M. (1977) 'Affinity groups', in Woodcock, G. (ed.), *The Anarchist Reader*, London: Fontana, 173–4.

Bookchin, M. (1980) *Towards an Ecological Society*, Montreal: Black Rose Books.

Bookchin, M. (1982) *The Ecology of Freedom: the Emergence and Dissolution of Hierarchy*, Palo Alto: Cheshire Books.

Bookchin, M. (1986a) 'A green course', *Resurgence*, 115, 10–13.

Bookchin, M. (1986b) *The Modern Crisis*, Philadelphia: New Society Publishers.

Bookchin, M. (1987) 'Social ecology versus "deep ecology" a challenge for the ecology movement', *The Raven*, 1(3), 219–50.

Bookchin, M. (1989) 'New social movements: the anarchic dimension', in Goodway, D. (ed.), *For Anarchism: History, Theory and Practice*, London: Routledge, 259–74.

Bookchin, M. (1990) 'Ecologising the dialectic', in Clark, J. (ed.), *Renewing the Earth: the Promise of Social Ecology*, Basingstoke: Green Print, 202–19.

Bookchin, M. (1992) 'Ecology as a dismal science', *Green Line*, 96, 11–12.

Bookchin, M. and Foreman, D. (1991) *Defending the Earth: a Debate between Murray Bookchin and Dave Foreman*, Montreal: Black Rose Books.

Bottomore, T., Harris, L., Kiernan, V., Miliband, R. (eds) (1983) *A Dictionary of Marxist Thought*, Oxford: Blackwell.

Bramwell, A. (1989) *Ecology in the Twentieth Century: a History*, London: Yale University Press.

Brown, L. Susan (1988) 'Anarchism, existentialism and human nature', *The Raven*, 2(1), 49–60.

Buick, A. (1987) 'Socialism and calculation', *Socialist Standard*, 83, 1000, 466–7.

Buick, A. (1990) 'A market by the way: the economics of nowhere', in Coleman, S. and O'Sullivan, P. (eds), *Wm Morris and News from Nowhere: a Vision of our time*, Bideford: Green Books, 151–68.

Buick, A. (1992) 'The rich are still there', *Socialist Standard*, 89, 1052, April, 57–8.

Buick, A. and Crump, J. (1986) *State Capitalism: the Wages System under New Management*, London: Macmillan.

Bullock, A. and Stallybrass, O. (1977) *The Fontana Dictionary of Modern Thought*, London: Fontana.

Burgess, R. (1978) 'The concept of nature in geography and Marxism', *Antipode*, 10(2), 1–11.

Button, J. (1988) *A Dictionary of Green Ideas*, London: Routledge.

Byrne, D. (1985) 'Just hol on a minute there: a rejection of André Gorz's "Farewell to the Working Class" ', *Capital and Class*, 24, 75–98.

Cahill, T. (1989) 'Cooperatives and anarchism: a contemporary perspective', in Goodway, D. (ed.), *For Anarchism: History, Theory and Practice*, London: Routledge, 235–55.

Callenbach, E. (1978) *Ecotopia*, London: Pluto Press.

Callenbach, E. (1981) *Ecotopia Emerging*, Berkeley, California: Banyan Tree Books.

Callicott, J. B. (1985) 'Intrinsic value, quantum theory and environmental ethics', *Environmental Ethics*, 7, 257–75.

Campbell, A., Keen, C., Norman, G. and Oakeshott, R. (1977) *Worker-Owners: the Mondragon Achievement*, London: Anglo-German Foundation for the Study of Industrial Society.

Capra, F. (1982) *The Turning Point*, London: Wildwood House.

Carter, A. (1988) *Marx: a Radical Critique*, Brighton: Wheatsheaf.

Carter, A. (1989) 'Outline of an anarchist theory of history', in Goodway, D. (ed.), *For Anarchism: History, Theory and Practice*, London: Routledge, 176–97.

Carter, A. (1991) 'Towards a green political theory', paper presented to a workshop on Green Political Theory at the European Consortium for Political Research Joint Sessions, Essex University, March.

Castells, M. (1978) *City, Class and Power*, London: Macmillan.

Chodorkoff, D. (1990) 'Social ecology and community development', in Clark, J. (ed.), *Renewing the Earth: the Promise of Social Ecology*, 69–79.

City of Bradford Metropolitan Council (1989) *Green Light: Bradford's Strategy for the Environment*, Bradford: City of Bradford Metropolitan Council.

Clark, J. (1990a) 'A new philosophy for the green movement', in Clark, J. (ed.), *Renewing the Earth: the Promise of Social Ecology*, Basingstoke: Green Print, 1–5.

Clark, J. (1990b) 'What is social ecology', in Clark, J. (ed.), *Renewing the Earth: the Promise of Social Ecology*, 5–11.

Cohen, G. (1990) in 'Against the State: Marx and Freud', London: BBC Radio 4 programme.

Cole, K., Cameron, J. and Edwards, C. (1983) *Why Economists Disagree*, London: Longman.

Coleman, S. (1982a) 'William Morris' vision of socialism', recorded lecture given to the Socialist Party of Great Britain, London.

Coleman, S. (1982b) 'A Marxist critique of anarchism', recorded lecture given to the Socialist Party of Great Britain, London.

Coleman, S. (1984) 'Fear of freedom: the development of personality in capitalist society', recorded lecture given to the Socialist Party of Great Britain, London.

Coleman, S. (1990) 'How matters are managed', in Coleman, S. and O'Sullivan, P. (eds), *William Morris and News from Nowhere; a Vision for our Time*, Bideford: Green Books.

Coleman, S. and O'Sullivan, P. (1990) *William Morris and News from Nowhere: a Vision for our Time*, Bideford: Green Books.

Cook, D. (1992) 'Now is the time for all good men ...', *Guardian*, 25 February, 17.

Cook, I. (1990) 'Anarchistic alternatives: an introduction', *Contemporary Issues in Geography and Education*, 3(2), 9–21.

Cooter, W. S. (1978) 'Ecological dimensions of medieval agrarian systems', *Agricultural History*, 52, 438–77.

Cosgrove, D. (1990) 'Environmental thought and action: pre-modern and post-modern', *Transactions of the Institute of British Geographers*, 15(3), 344–58.

Coward, R. (1989) 'Natural Movements', BBC film in the 'Notes on the Margin' series.

Cranston, M. (1966) Introduction to his translation of John-Jacques Rousseau's (1743) *The Social Contract*, Harmondsworth: Penguin.

Crump, J. (1990) 'How the change came: News from Nowhere and revolution', in Coleman, S. and O'Sullivan, P. (eds), *Wm Morris and News from Nowhere: a Vision of our Time*, Bideford: Green Books, 57–73.

Cuff, E. C. and Payne, G. C. (eds) (1981) *Perspectives in Sociology*, London: Allen and Unwin.

Cullen, S. (1991) 'Anarchism and the problem of nationalism' *Freedom*, 52(15).

D'Arcy, J. (1970) 'The materialist conception of history', recorded lecture given to the Socialist Party of Great Britain, London.

D'Arcy, J. and Baritrop, R. (1975) 'Historical materialism', recorded lecture given to the Socialist Party of Great Britain, London.

Dauncey, G. (1988) *After the Crash*, Basingstoke: Green Print.

Dawkins, R. (1976) *The Selfish Gene*, Oxford: Oxford University Press.

Deleage, J. P. (1989) 'Eco-Marxist critique of political ecology', *Capitalism, Nature, Socialism*, 3, 15–31.

Desai, M. (1983) 'Capitalism', in Bottomore, T., *et al.*, *Dictionary of Marxist Thought*, Oxford: Blackwell.

Devall, W. and Sessions, G. (1985) *Deep Ecology: Living as if Nature Mattered*, Utah: Gibbs M. Smith.

Dobson, A. (1990) *Green Political Thought*, London: Unwin Hyman.

Doheny, J. R. (1991) 'Natural anarchism', *Freedom*, 52, 16.

Donnison, D. (1991) 'Sinking with the tide', *Guardian*, 21 August, 19.

Eckersley, R. (1992) *Environmentalism and Political Theory: Towards an Ecocentic Approach*, London: UCL Press.

Ehrlich, P. and Ehrlich, P. (1972) *Population, Resources, Environment*, San Francisco: Freeman.

Ehrlich, P. and Ehrlich, P. (1990) *The Population Bomb*, New York: Simon and Schuster.

Ekins, P. (ed.) (1986) *The Living Economy*, London: Routledge and Kegan Paul.

Ekins, P. (1992) *A New World Order*, London: Routledge.

Elkin, T. and McLaren, D. (1991) *Reviving the City; Towards Sustainable Urban Development*, London: Friends of the Earth.

Elkington, J. and Burke, T. (1987) *The Green Capitalists*, London: Gollancz.

Elliot, D. (1983) 'Making greens see red', *New Statesman*, 19 August, 14–15.

Elsom, D. (1992) *Atmospheric Pollution: a Global Problem*, second edition, Oxford: Blackwell.

Elton, B. (1989) *Stark*, London: Sphere.

Ely, J. (1990) 'Animism and anarchism', in Clark, J. (ed.), *Renewing the Earth: the Promise of Social Ecology*, Basingstoke: Green Print, 49–65.

Encel, S. (1990) 'Reflections on a visit to Mondragon', *Bulletin of the International Communal Studies Association*, 8, 4–6.

Engel, J. R. and Engel, J. G. (1990) 'The ethics of sustainable development', in Engel, J. R. and Engel, J. G. (eds), *Ethics of Environment and Development: Global Challenge, International Response*, London: Belhaven, 1–23.

Engels, F. (1872) 'Letter to Theodor Cuno', in Feuer, L. S. (ed.), *Marx and Engels: Basic Writings on Politics and Philosophy*, London: Fontana.

Ensensberger, H. (1974) 'A critique of political ecology', *New Left Review*, 84, 3–32.

Etchart, J. (1992) 'I killed you because you have no future', *Guardian*, 6 June, 23.

Faber, D. (1988) 'Imperialism and the crisis of nature in Central America', *Capitalism, Nature, Socialism*, 1, Fall, 39–45.

Faber, D. and O'Connor, J. (1989) 'The struggle for nature: environmental crisis and the crisis of environmentalism in the US', *Capitalism, Nature, Socialism*, 2, 12–39.

Fox, W. (1990) *Towards a Transpersonal Ecology*, London: Shambhala.

Francis, D. (1991) 'How to survive an attack of New Age ideology', *Green Line*, 87, 12–13.

Frankel, B. (1987) *The Post Industrial Utopians*, Cambridge: Polity Press.

Fry, C. (1975) 'Marxism and ecology', *Ecologist*, 6(9), 328–32.

Galbraith, J. K. (1991) 'Finding freedom in a world of poverty', *Guardian*, 27 August, 15.

German, L. (undated) *Why We Need a Revolutionary Party*, London: Socialist Workers' Party.

REFERENCES

Goldman, M. and O'Connor, J. (1988) 'Ideologies of environmental crisis: technology and its discontents', *Capitalism, Nature, Socialism*, 1(1), 91–106.

Goldsmith, E. (1978) 'The religion of a stable society' *Man-Environment Systems*, 8, 13–24.

Goldsmith, E. (1988) *The Great U-Turn*, Bideford: Green Books.

Goldsmith, E. (1990) 'The Uruguay round: gunboat diplomacy by another name', *Ecologist*, 20(6), 202–4.

Goldsmith, E. *et al.* (1972) 'A blueprint for survival', *Ecologist*, 2(1), 1–43.

Goodway, D. (1989) Introduction to Goodway, D. (ed.), *For Anarchism: History, Theory and Practice*, London: Routledge, 1–22.

Goodwin, B. (1982) *Using Political Ideas*, London: Wiley.

Goodwin, B. and Taylor, K. (1982) *The Politics of Utopia*, London: Hutchinson.

Gorz, A. (1982) *Farewell to the Working Classes: an Essay on Post-Industrial Socialism*, London: Pluto.

Gotts, N. (1992) 'Ghosts from the European past', *Green Line*, 97, 10.

Gould, P. (1988) *Early Green Politics: Back to Nature, Back to the Land and Socialism in Britain*, Brighton: Harvester Press.

Graham, R. (1989) 'The role of contract in anarchist theory', in Goodway, D. (ed.), *For Anarchism: History, Theory and Practice*, London: Routledge, 150–75.

Gregory, D. (1981) 'Regional geography', entry in Johnston, R. J. (ed.), *The Dictionary of Human Geography*, Oxford: Blackwell.

Griffiths, J. (1990) 'The collective unfairness of laissez faire', *Guardian*, 14 June.

Grossman, R. (1985) 'Environmentalists and the labor movement', *Socialist Review*, 15(4 and 5), 63–87.

Grundmann, R. (1991) *Marxism and Ecology*, Oxford: Clarendon Press.

Gudynas, E. (1990) 'The search for an ethic of sustainable development in Latin America', in Engel, J. R. and Engel, J. G. (eds), *Ethics of Environment: Global Challenge, International Response*, London: Belhaven, 139–49.

Guerin, D. (1989) 'Marxism and anarchism', in Goodway, D. (ed.), *For Anarchism: History, Theory and Practice*, London: Routledge, 109–25.

Guha, R. (1991) 'Lewis Mumford: the forgotten American environmentalist: an essay in rehabilitation', *Capitalism, Nature, Socialism*, 2(3), 67–91.

Habermas, J. (1983) 'Modernity: an incomplete project', in Foster, H. (ed.), *The Anti-Aesthetic: Essays on Postmodern Culture*, Port Townsend, Washington: Bay Press. Washington.

Haigh, M. (1988) 'Understanding "Chipko": the Himalayan people's movement for forest conservation', *International Journal for Environmental Studies*, 31, 99–110.

Hall, W. (1991) Review of Wall, D., *Getting There: Steps to a Green Society*, in *Capitalism, Nature, Socialism*, 2(3), 127–34.

Hallam, N. J. and Pepper, D. M. (1990) 'Feminism, anarchism and ecology: some connections', *The Raven*, 3(1), 46–55.

Hamer, M. (1987) *Wheels within Wheels*, London: Routledge and Kegan Paul.

Hardin, G. (1968) 'Tragedy of the commons', *Science*, 162, 1243–8.

Hardin, G. (1974) 'Living on a lifeboat', *Bioscience*, 24, 10.

Hardy, D. (1979) *Alternative Communities in Nineteenth Century England*, London: Longman.

Hardy, E. (1970) 'Marxian economics: chapters 1–4 of "Capital" ', recorded lecture given to the Socialist Party of Great Britain, London.

Hardy, E. (1977) 'Marxism since Marx', recorded lecture given to the Socialist Party of Great Britain, London.

Harper, C. (1987) *Anarchy: a Graphic Guide*, London: Camden Press.

Harrison, H. (1982) *Make Room! Make Room!*, Harmondsworth: Penguin.

Harvey, D. (1974) 'Population, resources and the ideology of science', *Economic Geography*, 50, 256–77.

Harvey, D. (1982) *The Limits to Capital*, Oxford: Blackwell.

Harvey, D. (1990) *The Condition of Postmodernity*, Cambridge: Polity.

Hay, P. R. (1988) 'Ecological values and Western political traditions; from anarchism to fascism', *Politics*, 8(2), 22–9.

Hebdige, R. (1989) 'After the masses', *Marxism Today*, January, 48–53.

Heilbroner, R. (1980) *Marxism, For and Against*, London: Norton.

Higgins, R. (1980) *The Seventh Enemy*, London: Pan.

Hoult, C. (1991) *Living Green: a Summer's Cycle around Green Britain*, Bideford: Green Books.

Howard, E. (1898) *Garden Cities of Tomorrow*, London: Faber and Faber.

Hulsberg, W. (1985) 'The greens at the crossroads', *New Left Review*, 152, 5–29.

Hunt, R. (undated) *The Natural Society: a Basis for Green Anarchism*, Oxford: EOA Books.

Ignatieff, M. (1986) 'Time to take new political bearings', *The Listener*, May, 16–17.

Jennings, T. (1990) 'Politics and the class struggle in the 1990s: liberation theory', *The Raven*, 3(3), 201–12.

Johnston, R. J. (1981) *The Dictionary of Human Geography*, Oxford: Blackwell.

Johnston, R. J. (1989) *Environmental Problems: Nature, Economy and State*, London: Belhaven.

Joll, J. (1979) *The Anarchists*, London: Methuen.

Kamenka, E. (1982a) 'Community and the socialist ideal', in Kamenka, E. (ed.), *Community as a Social Ideal*, London: Arnold, 1–26.

Kamenka, E. (1982b) 'The idea of community', in Kamenka, E. (ed.), *Community as a Social Ideal*, London: Arnold.

Kearney, R. (1986) *Modern Movements in European Philosophy*, Manchester: Manchester University Press.

Keleman, P. (1986) 'The politics of famine', *Contemporary Issues in Geography and Education*, 2(2), 14–29.

Khor Kok Peng, M. (1990) 'The Uruguay round and the third world', *Ecologist*, 20(6), 208–11.

Khozin, G. (1979) *The Biosphere and Politics*, Moscow: Progress Publishers.

Kimber, R. and Richardson, J. (eds) (1974) *Campaigning for the Environment*, London: Routledge and Kegan Paul.

Kohr, L. (1957) *The Breakdown of Nations*, London: Routledge and Kegan Paul.

Kropotkin, P. (1899) *Fields, Factories and Workshops Tomorrow*, London: Freedom Press.

Kropotkin, P. (1902) *Mutual Aid*, London: Freedom Press.

Lacey, A. R. (1986) *A Dictionary of Philosophy*, London, Routledge.

Laptev, I. (1990) 'Raising the biosphere to the noosphere', in Engel, J. R. and Engel, J. G. (eds), *Ethics of Environment and Development*, 117–26.

Lash, S. and Urry, J. (1987) *The End of Organised Capitalism*, Cambridge: Polity.

Lawrence, P. (1982) 'Marx and alienation', recorded lecture given to the Socialist Party of Great Britain, London.

Leopold, A. (1949) *A Sand County Almanack*, New York: Oxford University Press.

Lovejoy, A. (1974) *The Great Chain of Being*, Cambridge, Mass.: Harvard University Press.

Lovelock, J. (1989) *The Ages of Gaia: a Biography of our Living Earth*, Oxford: Oxford University Press.

Lowe, P. and Worboys, M. (1978) 'Ecology and the end of ideology', *Antipode*, 10(2), 12–21.

Lukes, S. (1973) *Individualism*, Oxford: Blackwell.

McKibben, W. (1990) *The Death of Nature*, Harmondsworth: Penguin.

Mangin, W. and Turner, J. (1969) 'Benavides and the Barriada movement', in Oliver, P. (ed.), *Shelter and Society*, New York: Barrie and Rockliff.

Marcovic, M. (1990) 'The development vision of socialist humanism' in Engel, J. R. and Engel, J. G. (eds), *Ethics of Environment and Development: Global Challenge, International Response*, London: Belhaven.

Marshall, P. (1989) 'Human nature and anarchism', in Goodway, D. (ed.), *For Anarchism: History, Theory and Practice*, 127–49.

Martinez-Allier, J. (1990) *Ecological Economics: Energy, Environment and Society*, Oxford: Blackwell.

Marx, K. (1859) Preface of *A Contribution to the Critique of Political Economy*, excerpts in Feuer, L. (ed.), *Marx and Engels: Basic Writings*, London: Fontana, 84.

Marx, K. (1959) *Capital*, vol. 1, Moscow: Foreign Languages Publishing House.

Marx, K. and Engels, F. (1981) *The German Ideology*, New York: International Publishers.

Matley, I. (1966) 'The Marxist approach to the geographical environment', *Annals of the Association of American Geographers*, 56, 97–111.

Matley, I. (1982) 'Nature and society: the continuing Soviet debate', *Progress in Human Geography*, 6(3), 367–96.

Medvedev, Z. A. (1969) *The Rise and Fall of T. D. Lysenko*, New York: Columbia University Press.

Medvedev, Z. A. (1979) *Soviet Science*, Oxford: Oxford University Press.

Merrill, R. (1990) 'Reflections on science, technology and the biological paradigm' in Clark, J. (ed.), *Renewing the Earth: the Promise of Social Ecology*, 33–48.

Milbrath, L. W. (1989) *Envisioning a Stable Society: Learning our Way Out*, Albany, NY: State University of New York Press.

Miliband, R. (1989) *Divided Societies: Class Struggle in Contemporary Capitalism*, Oxford: Oxford University Press.

Moore, S. (1990) 'Green light spells danger', *Guardian*, 3 May, 17.

Morris, M. (1992) 'The man in the mirror: David Harvey's "condition" of postmodernity', *Theory, Culture and Society*, 9, 253–79.

Morris, Wm (1885) 'Useful work versus useless toil', Socialist League pamphlet, in *Collected Works*, Penguin edition, 117–36.

Morris, Wm (1887) 'How we live and how we might live' (originally published in *Commonweal*), *Collected Works*, XXIII, 3–26.

Morris, Wm (1890) *News from Nowhere*, London: Routledge and Kegan Paul, 1970 edition.

Mosse, G. L. (1982) 'Nationalism, fascism and the radical right', in Kamenka, E. (ed.), *Community as a Social Ideal*, London: Arnold, 27–42.

Mumford, L. (1934) *The Future of Technics and Civilisation* (second half of *Technics and Civilisation*), London: Freedom Press

Mundey, J. (1981) *Green Bans and Beyond*, Sydney: Angus and Robertson.

Murray, F. (1987) 'Flexible specialisation in the "Third Italy" ', *Capital and Class*, 33, 84–95.

Neville, P. (1990) 'Class', *The Raven*, 3(3), 230–5.

O'Connor, J. (1988) 'Capitalism, Nature, Socialism: a theoretical introduction', *Capitalism, Nature, Socialism*, 1(1), 11–38.

O'Connor, J. (1989) 'Political economy and ecology of socialism and capitalism', *Capitalism, Nature, Socialism*, 3, 93–106.

O'Connor, J. (1991a) 'Is sustainable capitalism possible?', in *Conference Papers by*

James O'Connor, Santa Cruz: Capitalism, Nature, Socialism/Centre for Ecological Socialism Pamphlet No. 1, 11–15.

O'Connor, J. (1991b) ' "External natural" conditions of production, the state, and political strategy for ecology movements', in *Conference Papers by James O'Connor*, Santa Cruz: Capitalism, Nature, Socialism/Centre for Ecological Socialism Pamphlet No. 1, 23–8.

O'Connor, J. (1991c) 'Socialism and ecology', in *Conference Papers by James O'Connor*, Santa Cruz: Capitalism, Nature, Socialism/Centre for Ecological Socialism Pamphlet No. 1, 29–40.

O'Connor, J. and Orton, D. (1991) 'Socialist biocentrism', *Capitalism, Nature, Socialism*, 2(3), 93–100.

Oldroyd, D. (1980) *Darwinian Impacts*, Milton Keynes: Open University Press.

Omari, C. K. (1990) 'Traditional African land ethics', in Engel, J. R. and Engel, J. G. (eds), *Ethics of Environment and Development*, London: Belhaven, 167–75.

Omo-Fadaka, J. (1990) 'Communalism: the moral factor in African development', in Engel, J. R. and Engel, J. G. (eds), *Ethics of Environment and Development*, 176–82.

O'Neil, J. (1988) 'Market socialism and information: a reformulation of a Marxian objection to the market', *Social Philosophy and Policy*, 6(2), 200–10.

Open University (1986) *Cooperative Working: Introducing Worker Cooperatives*, Milton Keynes: Open University Press.

O'Riordan, T. (1981) *Environmentalism*, second edition, London: Pion.

O'Riordan, T. (1989) 'The challenge for environmentalism', in Peet, R. and Thrift, N. (eds), *New Models in Geography*, London: Unwin Hyman, 77–102.

O'Sullivan, P. (1990) 'The ending of the journey: Wm Morris, News from Nowhere and ecology, in Coleman, S. and O'Sullivan, P. (eds), *Wm Morris and News from Nowhere: a Vision for Our Time*, Bideford: Green Books, 169–81.

Pallister, D. and Norton-Taylor, R. (1992) 'Open arms welcome officer class to defence industry', *Guardian*, 10 September, 4.

Papworth, J. (1990) 'The fourth world: the world of small units', *Noah's Ark*, 2, Leopold Kohr Extra, unpaginated.

Parsons, H. L. (1977) *Marx and Engels on Ecology*, London: Greenwood.

Parsons, J. (1990) 'Too many people: too much taboo?', *New Ground*, Spring, 10.

Paterson, T. (1989) *The Green Conservative: a Manifesto for the Environment*, A Bow Paper, London: Bow Group.

Paxman, J. (1991) *Friends in High Places: who runs Britain?*, Harmondsworth: Penguin.

Pearce, D., Markandya, A. and Barbier, E. (1989) *Blueprint for a Green Economy*, London: Earthscan.

Peet, R. (1985) 'The social origins of environmental determinism', *Annals of the Association of American Geographers*, 75(3), 309–33.

Peet, R. (1991) *Global Capitalism: Theories of Societal Development*, London: Routledge.

Pena, D. (1992) 'The "Brown" and the "Green": Chicanos and environmental politics in the Upper Rio Grande', *Capitalism, Nature, Socialism*, 3(1), 79–103.

Pepper, D. M. (1980) 'Environmentalism, the "lifeboat ethic" and anti-airport protest', *Area*, 12(3), 177–82.

Pepper, D. M. (1984) *Roots of Modern Environmentalism*, London: Routledge.

Pepper, D. M. (1988) 'The geography and landscapes of an anarchist Britain', *The Raven*, 1(4), 339–50.

Pepper, D. M. (1991) *Communes and the Green Vision: Counterculture, Lifestyle and the New Age*, Basingstoke: Green Print.

REFERENCES

Pepper, D. and Hallam, N. (1988) 'How can the green movement break through?', *Green Line*, 60, March, 10–11.

Pepper, D. M. and Hallam, N. J. (1989) 'The Findhorn tendency', *New Ground*, 20, 18–20.

Perelman, M. (1979) 'Marx, Malthus and the concept of natural resource scarcity', *Antipode*, 11(2), 80–9.

Piercey, M. (1979) *Woman on the Edge of Time*, London: The Women's Press.

Porritt, J. (1984) *Seeing Green*, Oxford: Blackwell.

Porritt, J. and Winner, D. (1988) *The Coming of the Greens*, London: Fontana.

Pringel, P. and Spigelman, J. (1983) *The Nuclear Barons*, London: Sphere Books.

Purchase, G. (1990) *Anarchist Society and its Practical Realisation*, San Francisco: See Sharp Press.

Purchase, G. (1992) *Social Ecology, Anarchism and Trades Unionism*, Rebel Worker Pamphlet No. 11, Sydney: ASF–IWA (Anarcho-Syndicalist Federation–International Workers' Association, PO Box 92, Broadway, NSW 2007).

Quaini, M. (1982) *Geography and Marxism*, Oxford: Blackwell.

Raghavan, C. (1990) 'Recolonisation: GATT in its historical context', *Ecologist*, 20(6), 205–7.

Redclift, M. (1986) 'Redefining the environmental "crisis" in the South' in Weston, J. (ed.), *Red and Green: a New Politics of the Environment*, London: Pluto Press, 80–101.

Redclift, M. (1987) *Sustainable Development*, London: Routledge.

Redmond, J. (1983) Editor's introduction to Wm Morris, *News from Nowhere*, London: Routledge and Kegan Paul edition.

Rheingold, H. (1991) *Virtual Reality*, London: Secker and Warburg, excerpted in the *Guardian*, 26 and 27 August.

Richards, F. (1989) 'Can capitalism go green?', *Living Marxism*, 4, 18–23.

Rigby, A. (1974) *Alternative Realities: a Study of Communes and their Members*, London: Routledge and Kegan Paul.

Rigby, A. (1990) 'Lessons from anarchist communes' *Contemporary Issues in Geography and Education*, 3(3), 52–62.

Ritchie, M. (1990) 'GATT, agriculture and the environment: the US double zero plan', *Ecologist*, 20(6), 214–20.

Robertson, J. (1983) *The Sane Alternative: a Choice of Futures*, Ironbridge, Salop: J. Robertson.

Robertson, J. (1990) *Future Wealth*, London: Cassell.

Rooum, D. (1987) 'Anarchism and selfishness', *The Raven*, 1(3), 251–9.

Rooum, D. (1992) 'Individualist anarchism is class struggle anarchism', *Freedom*, 53(2), 25 January, 2.

Rose, S., Kamin, R. and Lewontin, L. (1984) *Not in our Genes*, Harmondsworth: Penguin.

Rostow, W. W. (1960) *The Stages of Economic Growth: a Non-communist Manifesto*, Cambridge: Cambridge University Press.

Roszak, T. (1970) *The Making of a Counter Culture*, London: Faber.

Roszak, T. (1979) *Person/Planet*, London: Gollancz.

'RSW' (1990) 'Marxist-communism *has* failed' *Freedom*, 51(12), 16 June, 6.

Russell, B. (1918) *Roads to Freedom*, third edition 1948, London: Unwin, 1977.

Russell, B. (1946) *History of Western Philosophy*, London: Unwin.

Ryle, M. (1988) *Ecology and Socialism*, London: Radius.

Sabel, C. (1982) *Work and Politics*, Cambridge: Cambridge University Press.

Sale, K. (1985) *Dwellers in the Land: the Bioregional Vision*, San Francisco: Sierra Club.

Sampson, A. (1984) *The Changing Anatomy of Britain*, New York: Vintage Books.

Sarkar, S. (1983) 'Marxism and productive forces: a critique', *Alternatives*, IX, 145–76.

Sayer, A. (1983) 'Notes on geography and the relationship between people and nature', in Cannon, T., Forbes, M. and Mackie, J. (eds), *Society and Nature*, London: Union of Socialist Geographers.

Sayer, A. (1992) 'Radical geography and Marxist political economy: towards a re-evaluation', *Progress in Human Geography*, 16(3), 343–60.

Sayer, A. and Walker, R. (1992) *The New Social Economy: Reworking the Division of Labour*, Oxford: Blackwell.

Schmidt, A. (1971) *The Concept of Nature in Marx*, London: New Left Books.

Schnaiberg, A. (1980) *The Environment: from Surplus to Scarcity*, New York: Blackwell.

Schumacher, E. F. (1973) *Small is Beautiful: Economics as if People Really Mattered*, London: Abacus.

Schumacher, E. F. (1980) *Good Work*, London: Abacus.

Schwarz, W. and Schwarz, D. (1987) *Breaking Through: Theory and Practice of Wholistic Living*, Bideford: Green Books.

Scott, A. (1990) *Ideology and the New Social Movements*, London: Unwin Hyman.

Scruton, R. (1980) *The Meaning of Conservatism*, London: Macmillan.

Seabrook, J. (1985) *Landscapes of Distress*, Oxford: Blackwell.

Seabrook, J. (1986) 'The inner city environment: making the connections', in Weston, J. (ed.), *Red and Green: The New Politics of the Environment*, London: Pluto Press, 102–12.

Seabrook, J. (1989) *The Race for Riches*, Basingstoke: Green Print.

Seabrook, J. (1990) *The Myth of the Market*, Bideford: Green Books.

Sekelj, L. (1990) 'Has anarcho-communism a future?', *The Raven*, 3(1), 15–25.

Sennett, R. (1978) *The Fall of Public Man*, London: Random House.

Shenfield, B. (1971) *Company Boards*, London: Allen and Unwin.

Short, J. R. (1991) *Imagined Country: Society, Culture and Environment*, London: Routledge.

Simcock, J. (1991) 'Anarchism and cities', *Freedom*, 52(22), 5–6.

Simon, J. and Kahn, H. (1984) *The Resourceful Earth: a Response to Global 2000*, Oxford: Blackwell.

Sjoo, M. (1992) 'Questioning the new age movement', *Green Line*, 96, 13–14.

Slapper, C. (1983) 'Ending problems of world hunger', recorded lecture given to the Socialist Party of Great Britain, London.

Smith, A. (1987) 'Jobs and the environment: the role of local authorities', in Minhay, C. and Weston, J. (eds), *The Future of Work; Jobs in the Environment*, Oxford: Oxford Polytechnic Department of Town Planning Working Paper 103, 73–85.

Smith, N. (1984) *Uneven Development*, Oxford: Blackwell.

Smith, N. and O'Keefe, P. (1980) 'Geography, Marx and the concept of nature', *Antipode*, 12(2), 30–9.

SPGB (Socialist Party of Great Britain) (1983) 'Conservation versus democratic revolution', recorded debate with the Ecology Party, London.

SPGB (1987) *Socialism as a Practical Alternative*, London: SPGB.

SPGB (1990) *Ecology and Socialism*, London: SPGB.

Spoehr, A. (1967) 'Cultural differences in the interpretation of natural resources', in Thomas, W. (ed.), *Man's Role in Changing the Face of the Earth*, 93–101.

Steward, F. (1989) 'New times; green times', *Marxism Today*, March, 14–17.

Stoneham, C. (1972) *The Unviability of Capitalism*.

Stretton, H. (1976) *Capitalism, Socialism and the Environment*, Cambridge: Cambridge University Press.

REFERENCES

Taffe, P. (1990) *Militant: What we Stand For*, London: Militant Publications.

Thomas, K. (1983) *Man and the Natural World*, London: Allen Lane.

Thompson, J. L. (1983) 'Preservation of wilderness and the good life' in Elliot, R. and Gare, A. (eds), *Environmental Philosophy*, Milton Keynes: Open University Press.

Tonnies, F. (1887) *Community and Society*, New York: Harper and Row, 1963.

Touraine, A. (1986) 'The end of the old political passions', interviewed with R. Darhendorf by Michael Ignatieff in *The Listener*, 10 April, 16–17.

Ullrich, O. (1979) *World Standard: in the Blind Alley of the Industrial System*, Berlin: Ratbuch Verlag.

UN (1987) World Commission on Environment and Development, *Our Common Future* (The Brundtland Report), Oxford: Oxford University Press.

Vaillancourt, J.-G. (1992) 'Marxism and ecology: more Benedictine than Franciscan', *Capitalism, Nature, Socialism*, 3(1), 19–35.

Vallely, P. (1985) 'How Mengistu hammered the peasants', *The Times*, 1 March.

Van der Weyer, R. (1986) *Wickwyn: a Vision of the Future*, London: SPCK.

Vidal, J. (1992) 'The Wolverhampton wanderers', *Guardian*, 12 September, 22.

Vidal, J. and Chaterjee, P. (1992) 'All the difference in the world', *Guardian*, 10 April, 19.

Vogel, S. (1988) 'Marx and alienation from nature', *Social Theory and Practice*, 14(3), 367–88.

Wainwright, H. and Elliot, D. (1982) *The Lucas Plan: a New Trade Unionism in the Making*, London: Alison and Busby.

Walford, G. (1990) 'Class politics: an exhausted myth', *The Raven*, 3(3), 225–30.

Wall, D. (1990) *Getting There: Steps to a Greener Society*, Basingstoke: Green Print.

Wall, D. (1991) 'Goodbye to the Green Party?', *Green Line*, 89, 14–15.

Walters, M. (1980) 'Is human nature a barrier to socialism?' Recorded lecture given to the Socialist Party of Great Britain, London.

Ward, C. (1990) 'An anarchist looks at urban life', *Contemporary Issues in Geography and Education*, 3(2), 80–93.

Ward, C. (1991) 'Mondragon unvisited', *New Statesman*, 1 February.

Ward, C. and Hardy, D. (1986) *Goodnight Campers: the history of the British Holiday Camp*, London: Mansell Publishing.

Warnock, M. (1970) *Existentialism*, Oxford: Oxford University Press.

Watkinson, R. (1990) 'The obstinate refusers: work in News from Nowhere', in Coleman, S. and O'Sullivan, P. (eds), *News from Nowhere: a Vision for our Time*, Bideford: Green Books, 91–106.

Watson, R. (1983) 'A critique of non-anthropocentric biocentrism', *Environmental Ethics*, 3, 245–56.

Webster, F. and Robins, K. (1981) 'Information technology: futurism, corporations and the state', in Miliband, R. and Saville, J. (eds), *The Socialist Register 1981*, London: Merlin, 247–69.

Weston, D. (1992) 'Money and worth', in Button, J. and Bloom, W. (eds), *The Seeker's Guide: a New Age Resource Book*, London: Aquarian Press, 211.

Weston, J. (1986) 'The greens, "nature" and the social environment', in Weston, J. (ed.), *Red and Green: the New Politics of the Environment*, London: Pluto Press, 1–29.

White, L. (1967) 'The historical roots of our ecologic crisis', *Science*, 155, 1203–7.

Wingrove, I. (1991) 'Unions and disunity', *Green Line*, 87, 9.

Winner, L. (1986) *The Whale and the Reactor*, Chicago: University of Chicago Press.

Williams, R. (1983) 'Culture', in MacLellan, D. (ed), *Marx: the First 100 Years*, Oxford: OUP/Fontana, 15–56.

Wood, A. (1988) *Greentown: a Case Study of a Proposed Alternative Community*, Open

University Energy and Environment Research Unit Occasional Paper No. 57, Milton Keynes: Open University.

Woodcock, G. (1975) *Anarchism*, Harmondsworth. Penguin.

Woodcock, G. (ed.) (1977) *The Anarchist Reader*, London: Fontana.

Worsthorne, P. (1984) 'Inequality's role in democracy', *Sunday Telegraph*, 26 February.

Wylde, A. B. (1901) *Modern Abyssinia*, London: Methuen.

Yen, J. (1990a) 'Class, power, and class consciousness: an anarchist model', *The Raven*, 3(3), 195–200.

Yen, J. (1990b) 'Social thought and ideology', *The Raven*, 3(3), 263–86.

Young, J. (1990) *Post-Environmentalism*, London: Belhaven.

INDEX